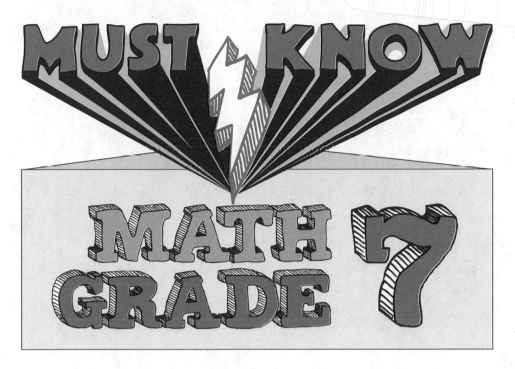

MUST KNOW
MATH GRADE 7

Wendy Hanks

Mc
Graw
Hill

New York Chicago San Francisco Athens London Madrid
Mexico City Milan New Delhi Singapore Sydney Toronto

1 2 3 4 5 6 7 8 9   LCR   25 24 23 22 21 20

ISBN   978-1-260-46690-4
MHID   1-260-46690-6

e-ISBN   978-1-260-46691-1
e-MHID   1-260-46691-4

Interior design by Steve Straus of Think Book Works.
Cover and letter art by Kate Rutter.

I would like to dedicate this book to my son, Noah Hanks, who was in a 7th grade math class while I wrote this book. He was an invaluable source of inspiration and motivation and kept me focused on the concepts you really must know. Thank you for your patience and your help. I would also like to thank Garret Lemoi for supporting me and my writing. I look forward to many more years of working with you.

Contents

Introduction

Welcome to your new math book! Let us try to explain why we believe you've made the right choice. You've probably had your fill of books asking you to memorize lots of terms (such as in school). This book isn't going to do that—although you're welcome to memorize anything you take an interest in. You may also have found that a lot of books make a lot of promises about all the things you'll be able to accomplish by the time you reach the end of a given chapter. In the process, those books can make you feel as though you missed out on the building blocks that you actually need to master those goals.

With *Must Know Math Grade 7,* we've taken a different approach. When you start a new chapter, right off the bat you will immediately see one or more **must know** ideas. These are the essential concepts behind what you are going to study, and they will form the foundation of what you will learn throughout the chapter. With these **must know** ideas, you will have what you need to hold it together as you study, and they will be your guide as you make your way through each chapter.

To build on this foundation, you will find easy-to-follow discussions of the topic at hand, accompanied by comprehensive examples that show you how to apply what you're learning to solving typical 7th-grade math questions. Each chapter ends with review questions—350+ throughout the book—designed to instill confidence as you practice your new skills.

This book has other features that will help you on this math journey of yours. It has a number of sidebars that will either provide helpful information or just serve as a quick break from your studies. The **BTW** sidebars ("by the way") point out important information, as well as tell you what to be careful about math-wise. Every once in a while, an 🌐 **IRL** sidebar ("in real life") will tell you what you're studying has to do with the real world; other IRLs may just be interesting factoids.

In addition, this book is accompanied by a flashcard app that will give you the ability to test yourself at any time. The app includes 100-plus "flashcards" with a review question on one "side" and the answer on the other. You can either work through the flashcards by themselves or use them alongside the book. To find out where to get the app and how to use it, go to the next section, "The Flashcard App."

We also wanted to introduce you to your guide throughout this book. We've had the pleasure of working on a number of projects with Wendy Hanks but are glad to, finally, have the opportunity to work with her on a project starting from scratch. Wendy has a clear idea what you should get out of a math class in 7th grade and has developed strategies to help you get there. She also understands the kinds of pitfalls that students can fall into and will help you solve those difficulties. In this book, Wendy applies that experience both to showing you the most effective way to learn a given concept and how to extricate yourself from any trouble you may have gotten into. She will be a trustworthy guide as you expand your math knowledge and develop new skills.

Before we leave you to Wendy's sure-footed guidance, let us give you one piece of advice. While we know that saying something is "the *worst*" is a cliché, if anything *is* the worst in the math you'll cover in 7th grade, it could be algebraic equations. Let the author introduce you to equations and show you how to work confidently with them. Take our word for it: learning how to handle algebraic equations will leave you in good stead for the rest of your math career—and in the real world, too.

Good luck with your studies!

The Editors at McGraw Hill

The Flashcard App

This book features a bonus flashcard app. It will help you test yourself on what you've learned as you make your way through the book (or in and out of it). It includes 100-plus "flashcards," both "front" and "back." It gives you two options as to how to use it. You can jump right into the app and start from any point that you want. Or you can take advantage of the handy QR codes at the end of each chapter in the book; they will take you directly to the flashcards related to what you're studying at the moment.

To take advantage of this bonus feature, follow these easy steps:

Search for **McGraw Hill Must Know** App from either Google Play or the App Store.

↓

Download the app to your smartphone or tablet.

↓

Once you've got the app, you can use it in either of two ways.

↙   ↘

| Just open the app and you're ready to go. | Use your phone's QR Code reader to scan any of the book's QR codes. |
| You can start at the beginning, or select any of the chapters listed. | You'll be taken directly to the flashcards that match your chapter of choice. |

↘   ↙

Get ready to test your math knowledge!

Author's Note

This book continues your study of many topics you learned in grade school, but my real goal is to begin preparing you for high school-level math. In 7th grade, we concentrate on ratios and rates, especially unit rate and graphing rates. You will also learn more about data presentations, measures of central tendency, and probability. We will expand your knowledge of equations and inequalities and of geometry.

As a teacher, I like to focus on real-world applications for mathematics. I want you to see how learning these concepts can help you as you move into your adult life. When you are done with this book, you will know how to calculate discounts, taxes, tips, and interest. You'll be a terrific shopper! I never liked classes in which I could not figure out when or how I would ever use what we were learning. I hope you will find this book to be a practical, useful tool.

Many example questions throughout the chapter walk you through, step by step, how to find the answer to the question. If you think you already know how to answer a question, feel free to try it on your own before you read the explanation. But do read the explanation: you might learn a shortcut or a bit of information that you didn't already know that can help you with similar questions. At the end of each chapter, there are many more practice questions for you to attempt on your own. Answers and explanations for those appear at the end of the book. There is also an app with flashcard questions for each chapter of the book, so you can even practice on the go.

Remember that learning math is a process. You won't always know the answers right away. Some concepts take patience to learn and practice to master. This is your book, so use it. Write notes for yourself, highlight important ideas. Work the questions in a separate notebook so that you can rework them later for more practice. If you make a mistake, don't worry about it. Mistakes are part of the learning process. Don't give up. Read the explanation and try again. When you are done with a chapter, look back over it and ask yourself if you feel totally comfortable with all the topics in the chapter. Reread the **must know** concepts from the beginning of the chapter to be sure that you do, in fact, know them. I hope you enjoy conquering 7th grade math!

Rational Number Properties

MUST ⚡ KNOW

⚡ A rational number is any number that can be written as a fraction with integers in the numerator and denominator. This includes all integers and both finite and repeating decimals.

⚡ The order of operations is: parentheses, exponents, multiplication and division, addition and subtraction. PEMDAS!

⚡ The absolute value of a number is its positive distance from zero on the number line, whether the number is positive or negative.

ou may remember that a **whole number** is like a counting number. The set of whole numbers starts at 0 and goes on forever. The set looks like this:

$$\{0, 1, 2, 3, 4, 5, 6, 7, 8, 9, 10 \dots\}$$

The set of whole numbers does not contain any negative numbers, fractions, or decimals.

Integers are the set of whole numbers, plus whole negative numbers. The set of integers does not include fractions or decimals either. The set of integers looks like this:

$$\{-5, -4, -3, -2, -1, 0, 1, 2, 3, 4, 5, \dots\}$$

A **rational number** is any number that can be written as a fraction with integers in the numerator and denominator. This includes all integers and both repeating and finite decimals. We will be working with rational numbers throughout the book.

Our goal for this chapter is to make sure you are up to speed on working with whole numbers and integers. We will also review number properties and the order of operations.

Adding and Subtracting Whole Numbers

To add whole numbers, line up the addends (the numbers you are adding) by place value. Add each place value, starting with the ones place. If the total for a place value has two digits, carry the first digit to the next column.

EXAMPLE

▶
$$\begin{array}{r} 748 \\ + 452 \\ \hline \end{array}$$

▶ Since, $8 + 2 = 10$ write down the 0 and then carry the 1 over to the tens column.

$$
\begin{array}{r}
\overset{1}{748} \\
+\ 452 \\
\hline
0
\end{array}
$$

▶ Continue adding by place value column, carrying over when necessary.

$$
\begin{array}{r}
\overset{1\ 1}{748} \\
+\ 452 \\
\hline
1{,}200
\end{array}
$$

To subtract whole numbers, line up the numbers by place value. Subtract each place value number, starting with the ones place. If the top number is larger than the bottom number, you must regroup by borrowing from the next place value.

EXAMPLE

▶
$$
\begin{array}{r}
748 \\
-\ 452 \\
\end{array}
$$

▶ Subtract the ones column: $8 - 2 - 6$.

▶ In the tens column, since you cannot subtract 5 from 4, borrow from the hundreds column.

$$
\begin{array}{r}
\overset{6\,14}{7\!\!\!/48} \\
-\ 4\,52 \\
\hline
2\,96
\end{array}
$$

Multiplying and Dividing Whole Numbers

To multiply whole numbers, line up the numbers by place value. Multiply the digit in the ones place of the bottom number by each of the digits in the top number, beginning with the ones place. If the total for a place value has two digits, carry the first digit to the next column and add it on *after* you have multiplied that column. After you have multiplied by the bottom number's ones digit, move on to the bottom number's tens digit. Put a 0 as a placeholder in the ones place for the product. Then multiply the digit in the tens place of the bottom number by each of the digits in the top number. Continue through all of the place values, adding another 0 as a placeholder for each additional place value. When you have multiplied by each digit of the bottom number, add up all the partial products you found.

EXAMPLE

$$\begin{array}{r} 78 \\ \times\ 12 \\ \hline \end{array}$$

▶ Multiply 2 times 78. $2 \times 8 = 16$, so carry the 1. $2 \times 7 = 14$, then add the carried over 1 to get 15.

$$\begin{array}{r} {}^{1} \\ 78 \\ \times\ 12 \\ \hline 156 \end{array}$$

▶ Now put a 0 as a ones placeholder beneath the first partial product and multiply 1 times 78.

$$\begin{array}{r} 78 \\ \times\ 12 \\ \hline 156 \\ 780 \end{array}$$

▶ Add the two partial products to find the total product. $156 + 780 = 936$.

$$
\begin{array}{r}
78 \\
\times\ 12 \\
\hline
156 \\
780 \\
\hline
936
\end{array}
$$

To divide whole numbers, take the **dividend**, the number to be divided, and write it under a division bar. Put the **divisor**, the number you are dividing by, to the left of the division bar. Divide from left to right. When the remainder is less than the divisor, bring down the next place value's digit and keep dividing. When you run out of place value digits, if you still have a remainder, put a decimal point at the end of the dividend and add a 0. Add a decimal point to the **quotient** (your answer) as well. Keep adding 0s to the dividend as necessary until there is no remainder.

▶ $236 \div 20$

▶ Draw a division bar around 236 and write 20 to the left of it.

$$20\overline{)236}$$

▶ 20 goes into 23 once, with a remainder of 3.

$$
\begin{array}{r}
1 \\
20\overline{)236} \\
20 \\
\hline
3
\end{array}
$$

▶ Bring down the 6 and divide again.

$$
\begin{array}{r}
11 \\
20\overline{)236} \\
20\downarrow \\
\hline
36
\end{array}
$$

▶ 20 goes into 36 once, with a remainder of 16. Since the dividend has no more digits, add a decimal point and a 0 to keep dividing.

$$
\begin{array}{r}
11.8 \\
20{\overline{\smash{\big)}\,236.0}} \\
\underline{20} \\
36 \\
\underline{20} \\
16\,0 \\
\end{array}
$$

▶ Don't forget to add the decimal in your answer, too!

Number Properties

Numbers have four basic properties: commutative, associative, distributive, and identity. Knowing and using these properties makes our lives–and computations!–easier.

The **commutative property** states that we can add numbers in any order and that we can multiply numbers in any order. This allows us to rearrange the numbers to be added or multiplied in any way that is convenient.

$$1 + 6 + 7 = 14 \text{ is the same as } 6 + 7 + 1 = 14$$

$$2 \times 5 \times 3 = 30 \text{ is the same as } 3 \times 2 \times 5 = 30$$

EXAMPLE

▶ Matthew has to put 24 packs of playing cards into a box. He tries to put the packs in two layers with four rows of three, but they do not fit in the box. How can he rearrange the packs so that they fit into the box?

▶ He knows that $2 \times 4 \times 3 = 24$, but since that did not work, he thinks about other ways he can arrange the packs.

▶ He knows that numbers being multiplied can be put in any order, so he tries arranging the packs as $3 \times 2 \times 4 = 24$ in three layers with two rows of four. They still don't fit.

▶ He tries another arrangement: $4 \times 2 \times 3 = 24$ in four layers with two rows of three. This time, they finally fit into the box. Thanks, math!

The **associative property** states that three or more numbers to be added or multiplied can be grouped together in any way. This allows us to group the numbers in any way that is convenient.

$$(2 + 6) + 8 = 6 + (8 + 2) = 16$$

$$4 \times (7 \times 12) = (4 \times 12) \times 7 = 336$$

EXAMPLE

▶ Randy fills drink orders for sports teams at the concession stand and must add up the number of drinks needed very quickly. He gets several orders at a time, each written on a slip of paper. Here is one batch of orders he received for bottles of water: 13 bottles, 5 bottles, 6 bottles, 7 bottles, and 14 bottles. How can he quickly find the total number of water bottles he needs to fill these orders?

▶ If he looks at the order slips the way they were given to him, he would need to add $13 + 5 + 6 + 7 + 14$.

▶ That isn't an easy calculation to do in his head, so he rearranges the order of the slips into this: $(13 + 7) + (6 + 14) + 5$.

▶ He knows that $13 + 7$ is 20 and $6 + 14$ is 20, so he can quickly calculate $20 + 20 + 5 = 45$.

The **distributive property** states that the sum of two numbers times a third number is equal to the sum of each of the addends (the numbers to be added) times the third number.

$$3 \times (6 + 9) = (3 \times 6) + (3 \times 9) = 45$$

Any time we see something like this: $4(8 + 5)$, we have two options. We can add $(8 + 5)$ first and then multiply the sum (13) by 4 to get 52, or we can distribute the 4 to get $(4 \times 8) + (4 \times 5) = 32 + 20 + 52$. We can choose whichever way seems easier to us.

EXAMPLE

▶ Mr. Voelkle has to order supplies for his class to do a science project. Each student needs a pumpkin, a small bottle of vinegar, and a box of baking soda. Pumpkins are $5 each, a bottle of vinegar is $2, and a box of baking soda is $1. Mr. Voelkle has 23 students in his class. How much will the supplies cost?

▶ He could multiply the cost of each of the supplies by 23 students and then add them up, but that would require four separate calculations.

$$(5 \times 23) = 115$$
$$(2 \times 23) = 46$$
$$(1 \times 23) = 23$$
$$115 + 46 + 23 = 184$$

▶ It will be much faster to add the cost of the supplies for one student and then multiply that by 23 students.

$$(5 + 2 + 1) = 8$$
$$\downarrow$$
$$8 \times 23 = 184$$

The **identity property** states that any number plus 0 equals that number and that any number times 1 equals that number.

$$145 + 0 = 145$$

$$145 \times 1 = 145$$

Speaking of zero, we need to be aware of another property of zero. It is called the **multiplication property of zero**. It states that any number times 0 equals 0. Think about it: if you have none of something, it doesn't matter what the something is. You have nothing:

$$145 \times 0 = 0$$

$$23.56 \times 0 = 0$$

$$\frac{3}{8} \times 0 = 0$$

$$y^5 \times 0 = 0$$

$$z \times 0 = 0$$

Order of Operations

The **order of operations** tells us in what order to perform a series of calculations. Pay careful attention to the rules for order of operations because you will use them beyond just arithmetic; the algebraic equations you will do in high school and college also follow the same order of operations.

The first thing we need to do when evaluating an expression with multiple calculations is check for parentheses. We do any calculations inside the

BTW

Don't forget that in the order of operations, multiplication and division are done from left to right, whichever comes first. The same is true for addition and subtraction.

parentheses first. Then we calculate any numbers that have exponents. Then we perform any multiplication and division, whichever comes first from left to right. Finally, we perform any addition and subtraction, whichever comes first, from left to right.

Many people need help remembering the order of operations. If we use the first initial for each word in the order of operations—**P**arentheses, **E**xponents, **M**ultiplication, **D**ivision, **A**ddition, **S**ubtraction—we get **PEMDAS**, which is easier to remember. When I was in school, I learned a silly little saying that helps me remember PEMDAS: "Please Excuse My Dear Aunt Sally." You can also make up your own saying—the sillier the better. How about "Please Exercise My Dog After School" or "Put Every Monster Down And Scream"?

▶ What is $4(5 + 2) + 2$?

▶ First, do the operation in parentheses: $5 + 2 = 7$. Now we have $4(7) + 2$.

▶ Since there are no exponents, move on to multiplication and division. Make sure to do those in the order they appear from left to right. Since there is only a multiplication calculation, do that: $4 \times 7 = 28$.

▶ Now we have $28 + 2 = 30$.

That was a pretty straightforward question, but sometimes the order of operations can be quite tricky. In fact, people sometimes get so confused that they post questions on the Internet and get several different answers based on how other people remember (or misremember) the order of operations. Here is an example of the type of question that looks really easy but regularly starts arguments on the Internet.

EXAMPLE

▶ What is $24 \div 4(1 + 2)$?

▶ First, do the operation in parentheses: $1 + 2 = 3$. Now we have $24 \div 4(3)$.

▶ Since there are no exponents, move on to multiplication and division. Make sure to do those in the order they appear from left to right. Since there is a division calculation first, do that first: $24 \div 4 = 6$.

▶ Now we have $6(3) = 18$.

▶ Where many people go wrong is by doing the multiplication before the division. They calculate $24 \div 4(3)$ by doing $4(3)$ first and getting 12 and then dividing 24 by 12 to get 2. Of course, this is the wrong answer, so then their math-smart friends laugh at them. Now you know the correct order of operations—and can probably stump your parents!

Absolute Value

Absolute value refers to a number's distance from zero on the number line. It does not matter whether the number is positive or negative, since absolute value is a measure of distance. -2 and $+2$ have the same absolute value because they are both a distance of two away from zero on the number line. Absolute value is written using a pair of vertical bars, like this: $|-2| = 2$ or $|+2| = 2$.

EXAMPLE

▶ What is $|-73| + |-6|$?

▶ -73 is 73 away from 0 on the number line, so $|-73| = 73$.

▶ $|-6|$ is 6 away from 0, so $|-6| = 6$.

　$73 + 6 = 79$

IRL Absolute value is like calculating distance. If you walk two miles away from your house and then the same two miles back, you are at the same starting point, but you have walked a total of four miles: $|+2| + |-2| = 4$.

Adding and Subtracting with Negative Numbers

Positive numbers are to the right of zero on the number line. Negative numbers are to the left of zero on the number line. Zero itself is neither positive nor negative.

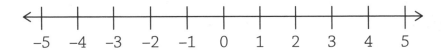

Each positive number has a negative counterpart, which is called the **inverse**. An **additive inverse** is the amount we need to add to a number to equal zero. The additive inverse of -4 is 4 because we have to add 4 to -4 to equal zero. The **property of additive inverses** says that when we add a number to its additive inverse, the result is zero. $-4 + 4 = 0$.

Addition and subtraction with negative numbers are more complicated than addition and subtraction with positive numbers, but we can use the number line and absolute value to help us.

Adding two negative numbers means we are starting on the left of zero on the number line, and then we are moving even farther to the left. To add $-2 + -6$, we start at -2 on the number line and then we move six more to the left, which puts us at -8. We can also add two negative numbers by adding the absolute value of the numbers and then putting a minus sign in front of the answer.

EXAMPLE

▶ What is $-2 + -6$?

▶ Find the absolute value of each number: $|-2| = 2$ and $|-6| = 6$.

▶ Now add the absolute values: $2 + 6 = 8$. Then put a minus sign on the answer, so $-2 + -6 = -8$.

To add one positive and one negative number, think of the number line and start at the positive number. Then move to the left the same number of spaces as the negative number. To find $2 + -6$, start at two on the number line and then move six spaces to the left, past zero, to -4. We can also add a positive and a negative number by finding the absolute value of the numbers and then subtracting the smaller one from the larger one. The answer will have the same sign as the original larger number.

EXAMPLE

▶ What is $2 + -6$?

▶ Find the absolute value of each number.

$$|2| = 2 \text{ and } |-6| = 6$$

▶ Subtract the smaller absolute value from the larger absolute value.

$$6 - 2 = 4$$

▶ For the answer, use the original sign of the larger absolute value. Since the larger absolute value is 6, use its original sign, which was a minus sign.

$$2 + -6 = -4$$

Subtracting a negative number gives us two minus signs together. You can think of this as adding the inverse of the negative number, which is positive. When we subtract a negative, we are really just adding!

▶ What is $2 - (-6)$?

▶ On the number line, start at 2 and then *add* 6, which means moving to the right along the number line, to 8. Subtracting (-6) is the same as adding 6.

$$2 - (-6) = 8$$

▶ What is $-2 - (-6)$? On the number line, start at -2 and then *add* 6, which means moving to the right along the number line, past 0, to 4.

$$-2 - (-6) = 4$$

Multiplying and Dividing with Negative Numbers

Multiplying and dividing negative numbers is much like multiplying and dividing positive numbers, but with two added rules to remember about the signs.

- When we multiply or divide two numbers with the *same sign* (either both positive or both negative), the answer is always *positive*.

- When we multiply or divide two numbers with *different signs* (one is positive and one is negative), the answer is always *negative*.

EXAMPLE

▶ What is $-2 \times (-6)$?

▶ Since both numbers have the same sign (both are negative), the answer will be positive. Simply multiply $2 \times 6 = 12$.

▶ What is $2 \times (-6)$?

▶ Since the numbers have different signs (one is positive and one is negative), the answer will be negative. Simply multiply $2 \times 6 = 12$ and then add a minus sign to get -12.

▶ What is $-8 \div (-4)$?

▶ Since both numbers have the same sign (both are negative), the answer will be positive. Simply divide $8 \div 4 = 2$.

▶ What is $8 \div (-4)$?

▶ Since the numbers have different signs (one is positive and one is negative), the answer will be negative. Simply divide $8 \div 4 = 2$ and then add a minus sign to get -2.

EXERCISES

EXERCISE 1-1

Add or subtract as indicated.

1.   41
    $$\underline{+\ 98}$$

2.   156
    $$\underline{+\ \ 75}$$

3.   9,872
    $$\underline{+\ 12,340}$$

4.   39
    $$\underline{-\ 23}$$

5.   482
    $$\underline{-\ 264}$$

6.   7,947
    $$\underline{-\ 3,608}$$

EXERCISE 1-2

Multiply or divide as indicated.

1.   27
    $$\underline{\times\ 46}$$

2. $\begin{array}{r} 352 \\ \times\ 105 \\ \hline \end{array}$

3. $4\overline{)183}$

4. $20\overline{)405}$

5. $48\overline{)6,294}$

EXERCISE 1-3

Perform the operations indicated. Remember to follow the order of operations.

1. $27 - 5 \times 3 + 1$

2. $(9 \div 3) + 4 \times 4$

3. $6(4 - 2) - 3 \div 3$

4. $5 \times 3(3 + 2)$

EXERCISE 1-4

Find the absolute value.

1. $|-46|$

2. $|506|$

3. $|-7|$

4. $|-3.5|$

5. $|0|$

EXERCISE 1-5

Add or subtract as indicated.

1. $-14 + -13$

2. $23 + -72$

3. $-8 - (-3)$

4. $6 - (-3)$

5. $-9 + 0$

EXERCISE 1-6

Multiply or divide as indicated.

1. $7(-7)$

2. $-4(-30)$

3. $60 \div (-5)$

4. $-18 \div (-3)$

5. $-3 \times 0 \times -12$

 **Fractions**

MUST ⚡ KNOW

⚡ A fraction shows how many equal parts are used out of a total number of equal parts.

⚡ To compare, add, or subtract fractions, find a common denominator or use the bow tie method.

⚡ To divide fractions, flip the second fraction over and multiply.

fraction is a part of a whole. I always like to think of fractions as pieces of a pizza. (I really like pizza!) If I cut a pizza into eight pieces and eat three pieces, I have eaten three out of the eight pieces. We express that as a fraction: $\dfrac{3}{8}$. We can also think of fractions as everything in between the integers on the number line. The fraction $\dfrac{1}{2}$ is halfway between 0 and 1 on the number line:

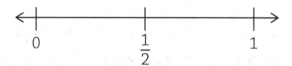

The number on the bottom of the fraction is called the **denominator**. It tells us the number of units that the whole is divided into. The whole pizza was cut into eight pieces, so the denominator is eight. The number on the top of the fraction is called the **numerator**. It tells us how many of the units there are. I ate three pieces of pizza, so the numerator is 3. Three out of eight total pieces $= \dfrac{3}{8}$. By the time you finish this chapter, you will know everything you need to know about ~~pizza~~ —I mean fractions. You will be able to compare, add, subtract, multiply, divide, and reduce fractions.

Improper Fractions and Mixed Numbers

If the numerator is less than the denominator, then the fraction is less than one. If I eat three out of eight pieces of pizza, that is less than eating the whole pizza. If I eat all eight pieces, then I ate the whole pizza. $\dfrac{8}{8} = 1$, the whole pizza, and I will probably have a stomachache.

If the numerator is greater than the denominator, then the fraction is greater than one. If I ate eleven pieces of pizza, then I ate more than just one whole eight-piece pizza. I ate that one whole pizza, plus three pieces of another eight-piece pizza, and I definitely have a stomachache. Eleven pieces out of an eight-piece pizza is written as $\frac{11}{8}$. This is called an **improper fraction**, since it is more than one.

We can change an improper fraction to a **mixed number** by showing how many whole items are used, plus the fraction of a whole that is left. If I ate eleven pieces from pizzas that are each cut into eight pieces, then I ate one whole pizza plus $\frac{3}{8}$ of another pizza, so we can write that as $1\frac{3}{8}$.

To change an improper fraction to a mixed number, divide the numerator by the denominator.

$$\frac{11}{8} = 11 \div 8 = 1 \text{ with a remainder of 3, so we get } 1\frac{3}{8}$$

To change a mixed number to an improper fraction, show the whole number as a fraction, and then add the numerators. If I ate $1\frac{3}{8}$ pizzas, then I ate all eight pieces of an eight-piece pizza, plus three more pieces of an eight-piece pizza, for a total of eleven pieces of eight-piece pizzas: $\frac{8}{8} + \frac{3}{8} = \frac{11}{8}$.

We can also multiply the denominator of the fraction part times the whole number, then add the numerator. That will give us the new numerator. The denominator stays the same.

$1\frac{3}{8}$ has 8 as the denominator, 3 as the numerator, and one as the whole number. Multiply the denominator (8) times the whole number (1): $8 \times 1 = 8$. Now add the numerator (3): $8 + 3 = 11$, which is the new numerator. The denominator stays the same (8), so we have $\frac{11}{8}$.

Equivalent Fractions

As you see, there is more than one way to write a fraction. You probably know that if you cut a pizza into eight slices and eat four, then you have eaten $\frac{1}{2}$ of the pizza. We can also say that you ate four out of the eight pieces, or $\frac{4}{8}$ of the pizza. This shows us that $\frac{4}{8}$ equals $\frac{1}{2}$. The same would be true if you cut that same pizza into six slices and ate three of them. You've eaten half of the pizza, so $\frac{3}{6}$ equals $\frac{1}{2}$. Since both $\frac{4}{8}$ and $\frac{3}{6}$ equal $\frac{1}{2}$, we say they are

equivalent fractions. This means they are equal: $\frac{4}{8} = \frac{3}{6} = \frac{1}{2}$.

If you eat the whole six-slice pizza, you have eaten $\frac{6}{6} = 1$ —a whole pizza! When the numerator and denominator are equal, the fraction is equal to one. We can take any fraction and multiply it by one without changing it because of the identity property for multiplication: any number times one equals that same number. When we do this with fractions, we do it with a fraction that equals one, such as $\frac{6}{6}$. This results in an equivalent fraction with a larger numerator and denominator.

▶ Change $\frac{1}{2}$ to a fraction with 12 in the denominator.

▶ How will we get from 2 in the denominator to 12? What times 2 equals 12? Six. We can multiply anything by 1 without changing the value of the original number, so let's multiply $\frac{1}{2}$ by one in the form of $\frac{6}{6}$:

$$\frac{1 \times 6}{2 \times 6} = \frac{6}{12}$$

$$\frac{6}{12} = \frac{1}{2}$$

We can increase the numerator and denominator of any fraction this way, by multiplying it with a fraction that equals one, without changing the actual value of the fraction.

Comparing and Ordering Fractions

One task you will need to be able to do is compare fractions. You may be asked to put a list of fractions in order from least to greatest, or you may be asked which of a group of fractions is largest.

Fractions with **like denominators** have exactly the same denominator. If the fractions have the same denominator, we can compare them easily by comparing the numerators: $\frac{3}{8} < \frac{5}{8}$. Eating three pieces of an eight-piece pizza is less than eating five pieces of an eight-piece pizza. (There I go with pizza again.)

If the fractions have different denominators, we cannot compare them that way. We need to find a **common denominator**, which is another way of saying a like denominator. To do that, think of the first few multiples for each denominator. When you find a multiple that they both have in common, that is a **common multiple** and can be used as the common denominator.

EXAMPLE

▶ Compare $\frac{3}{5}$ and $\frac{5}{6}$.

▶ Think of the first few multiples for the denominators, 5 and 6.

Multiples of 5: 5, 10, 15, 20, 25, 30 . . .
Multiples of 6: 6, 12, 18, 24, 30 . . .

▶ When we find a multiple that the denominators have in common, in this case 30, we can use that as the common denominator.

▶ Now convert each fraction into an equivalent fraction with 30 as the denominator. We do this by multiplying each fraction by one, which does not change the fraction, but we need to use one in the form of a fraction in which the numerator equals the denominator. Use whichever number for the numerator and denominator will make the denominator of the new fraction equal 30.

$$\frac{3}{5} \times \frac{6}{6} = \frac{18}{30} \text{ and } \frac{5}{6} \times \frac{5}{5} = \frac{25}{30}$$

▶ Now that the denominators are the same, we can compare the fractions:

$$\frac{18}{30} < \frac{25}{30}$$

If that seems complicated, there is another method that is usually faster. It is called the **bow tie method**. It is a quick and easy way to see which of two fractions is larger. Put the two fractions side by side and cross multiply the denominator of the first with the numerator of the second. Write the result on top of the second fraction. Then cross multiply the denominator of the second with the numerator of the first and write the result on top of the first fraction. Which number on top is larger? That's the larger fraction. Sound confusing? It's really quite simple to do. Let's look at the next example.

▶ Which is larger, $\frac{3}{8}$ or $\frac{4}{9}$?

▶ Write the fractions side by side and cross multiply.

$$\overset{27}{\frac{3}{8}} \diagdown\diagup \overset{32}{\frac{4}{9}}$$

▶ Compare the numbers in the boxes at the top. Since 32 is larger than 27, $\frac{4}{9}$ is larger than $\frac{3}{8}$.

If you are asked to compare two fractions and you use the bow tie method and the fractions end up with the same number on top, you will know that they are equivalent fractions.

EXAMPLE

▶ Compare $\frac{3}{4}$ and $\frac{9}{12}$.

▶ Write the fractions side by side and cross multiply.

$$\overset{36}{\frac{3}{4}} \diagdown\diagup \overset{36}{\frac{9}{12}}$$

▶ Compare the numbers in the boxes at the top. Since they are the same, the fractions are equivalent and $\frac{3}{4} = \frac{9}{12}$.

Another way to compare and order fractions is to convert them to decimals, and we will learn to do that in the next chapter.

To put a group of fractions in order, we can just repeat the process of comparing fractions. This takes more time, so if there are several fractions, it will probably be easier to use decimal conversion.

▶ Put these fractions in order, from least to greatest: $\frac{2}{5}, \frac{3}{7}, \frac{1}{3}$.

▶ Start with any two of the fractions. Let's just use the first two. Put them side by side and cross multiply.

$$\overset{14}{\frac{2}{5}} \times \overset{15}{\frac{3}{7}}$$

▶ Compare the numbers in the boxes at the top. Since 15 is larger than 14, $\frac{3}{7}$ is larger than $\frac{2}{5}$.

$$\frac{2}{5} < \frac{3}{7}$$

▶ Now let's compare the smaller one, $\frac{2}{5}$, to $\frac{1}{3}$.

$$\overset{6}{\frac{2}{5}} \times \overset{5}{\frac{1}{3}}$$

▶ Compare the numbers in the boxes at the top. Since 6 is larger than five, $\frac{2}{5}$ is larger than $\frac{1}{3}$. This gives us our complete order: $\frac{1}{3} < \frac{2}{5} < \frac{3}{7}$.

Adding and Subtracting Fractions

Adding and subtracting fractions with like denominators is fairly simple. Just add or subtract the numerators as indicated. The denominator stays the same. Think of it like eating pizza. Yes, really. If you eat two pieces of an eight-piece pizza and then you eat one more piece, you have eaten a total of three pieces of an eight-piece pizza.

▶ What is $\dfrac{2}{5}$ plus $\dfrac{1}{5}$?

▶ Since both fractions have a denominator of 5, that does not change. Simply add the numerators: $2 + 1 = 3$.

$$\frac{2}{5} + \frac{1}{5} = \frac{3}{5}$$

Subtraction works the same way if the fractions have like denominators. We leave the denominator alone and just do the subtraction with the numerators.

▶ What is $\dfrac{3}{7}$ minus $\dfrac{2}{7}$?

▶ Since both fractions have a denominator of 7, that does not change. Simply subtract the numerators: $3 - 2 = 1$.

$$\frac{3}{7} - \frac{2}{7} = \frac{1}{7}$$

Adding fractions with different denominators is more complicated. It would be difficult to figure out how much pizza you ate if you ate three pieces of a six-piece pizza, plus two slices of an eight-piece pizza. That wouldn't be just five pieces of pizza because the six-piece pizza probably has larger slices than the eight-piece pizza. We have to compare pizzas that are the same size. The same is true with fractions. We have to compare fractions that are on the same scale. We do that by finding a common denominator and converting the fractions to ones with that common denominator.

▶ What is $\dfrac{3}{5}$ plus $\dfrac{5}{6}$?

▶ Find a common denominator for 5 and 6. Let's use 30.

▶ Convert each fraction into a fraction with 30 as the denominator.

$$\frac{3}{5} \times \frac{6}{6} = \frac{18}{30} \text{ and } \frac{5}{6} \times \frac{5}{5} = \frac{25}{30}$$

▶ Now that the denominators are the same, we can add the numerators.

$$\frac{18}{30} + \frac{25}{30} = \frac{43}{30}$$

▶ Since that resulted in an improper fraction, we might need to change it to a mixed number. To express $\dfrac{43}{30}$ as a mixed number, we can divide 43 by 30 and get one with a remainder of 13, so the mixed number will be $1\dfrac{13}{30}$.

Let's try subtraction with unlike denominators. It is done the same way up until the last step. Find a common denominator and convert each fraction to one with that common denominator. Then instead of adding the numerators, subtract.

▶ What is $\dfrac{3}{4}$ minus $\dfrac{1}{6}$?

▶ Find a common denominator for 4 and 6. Let's use 12.

▶ Convert each fraction into a fraction with 12 as the denominator.

$$\frac{3}{4} \times \frac{3}{3} = \frac{9}{12} \text{ and } \frac{1}{6} \times \frac{2}{2} = \frac{2}{12}$$

▶ Now that the denominators are the same, we can subtract the numerators.

$$\frac{9}{12} - \frac{2}{12} = \frac{7}{12}$$

You may find it easier to use the bow tie method to add and subtract fractions. We do the same steps we did to compare two fractions with the bow tie: cross multiply and write the results on top of each fraction. Then we do the additional step of multiplying the denominators together. That will be the new denominator. For the new numerator, either add or subtract the numbers you wrote at the top of each fraction.

EXAMPLE

▶ What is $\frac{2}{5}$ plus $\frac{1}{3}$?

▶ First, do the cross multiplication blue that you would do to compare the fractions.

$$\overset{6}{\frac{2}{5}} \times \overset{5}{\frac{1}{3}}$$

▶ Now multiply the denominators.

$$\overset{6}{\frac{2}{5}} \times \overset{5}{\frac{1}{3}} = \frac{}{15}$$

▶ Now add the numbers you wrote at the top to find the new numerator.

$$\overset{6}{\frac{2}{5}} \times \overset{5}{\frac{1}{3}} = \frac{11}{15}$$

Multiplying Fractions

If you found adding and subtracting fractions challenging, then I have good news for you: multiplying fractions is much, much easier. All we have to do is multiply the numerators straight across to get the numerator of the product and then multiply the denominators straight across to get the denominator of the product.

EXAMPLE

▶ What is $\dfrac{2}{5}$ times $\dfrac{1}{3}$?

▶ Multiply the numerators and denominators straight across.

$$\frac{2 \times 1}{5 \times 3} = \frac{2}{15}$$

 IRL Being able to multiply fractions comes in handy when you are cooking. If your recipe serves four people, but you are only cooking for two, you will need to divide each ingredient in half. You may need to figure out what half of $\dfrac{1}{3}$ a pound is or half of $\dfrac{3}{4}$ a cup.

Fractions that are **reciprocals** look like the same fraction, but one is upside down. For example, $\dfrac{2}{5}$ and $\dfrac{5}{2}$ are reciprocals. When we multiply reciprocals, the product is always one.

EXAMPLE

▶ What is $\dfrac{2}{5}$ times $\dfrac{5}{2}$?

▶ Multiply the numerators and denominators straight across.

$$\frac{2 \times 5}{5 \times 2} = \frac{10}{10} = 1$$

▶ If you remember this rule, it can save you some time!

Sometimes when we multiply fractions, we end up with a very large product that may need to be **reduced** because we typically express fractions in their lowest form.

Reducing Fractions

To reduce a fraction down to its lowest form, we divide the fraction by a number that is a factor of both the numerator and the denominator. Remember how we can multiply anything by one without changing the value of the original number? We multiplied fractions by one (in fraction form) to change the denominator without changing the value of the fraction. Reducing is similar. We can divide any number by one without changing the value of the original number. We will do that with one in the form of a fraction.

EXAMPLE

▶ Reduce $\dfrac{6}{12}$.

▶ Since both 6 and 12 have 6 as a factor, we can divide the fraction by 1 in the form of $\dfrac{6}{6}$.

$$\frac{6 \div 6}{12 \div 6} = \frac{1}{2}$$

If you are dealing with a larger fraction and you do not immediately see a factor that the numerator and denominator have in common, remember that any even number is divisible by two, so if both the numerator and denominator are even numbers, start by dividing them by two.

▶ Reduce $\dfrac{72}{24}$.

▶ Since both 72 and 24 are even numbers, we can start by dividing them both by 2.

$$\dfrac{72 \div 2}{24 \div 2} = \dfrac{36}{12}$$

▶ Now you may see that both 36 and 12 are divisible by 12.

$$\dfrac{36 \div 12}{12 \div 12} = \dfrac{3}{1} = 3$$

▶ If not, just keep dividing by 2.

$$\dfrac{36 \div 2}{12 \div 2} = \dfrac{18}{6}$$

▶ One more time.

$$\dfrac{18 \div 2}{6 \div 2} = \dfrac{9}{3}$$

▶ Now the numerator is odd, so dividing by two won't work. When that happens, try dividing by 3.

$$\dfrac{9 \div 3}{3 \div 3} = \dfrac{3}{1} = 3$$

▶ Just keep reducing until you cannot reduce any more!

Some fractions cannot be reduced because they are already in their lowest form, but in general, if you can reduce a fraction, it will make your calculations easier.

To help us with reducing, here is a list of divisibility rules for the numbers 2 to 10, followed by some examples.

Divisibility Rules		
2	the last digit of the number is even	**62** or 1,568,94**6**
3	the sum of the digits is divisible by 3	73, 464: $7 + 3 + 4 + 6 + 4 = 24$, which is divisible by 3
4	the last two digits are divisible by 4	125,9**28:** Since 28 is divisible by 4, the whole number is divisible by 4
5	the last digit is a 0 or a 5	8**0** or 19,57**5**
6	the number is even *and* is divisible by 3	24**6**: the last digit is even *and* $2 + 4 + 6 = 12$, which is divisible by 3
8	the last three digits are divisible by 8	125,**928:** Since 928 is divisible by 8, the whole number is divisible by 8
9	the sum of the digits is divisible by 9	456,174: $4 + 5 + 6 + 1 + 7 + 4 = 27$, which is divisible by 9
10	the last digit is a 0	60**0** or 9,58**0**

These can help you reduce larger numbers that may not have an obvious factor.

EXAMPLE

▶ Reduce $\dfrac{543}{4,371}$ to its lowest form.

▶ This looks tough. The numerator and denominator are not even, so we can't reduce by a factor of 2, 4, 6, 8, or 10. Let's check 3. The rule for 3 is that if the digits add up to something divisible by 3, then the whole number is divisible by 3.

$$\frac{5+4+3=12}{4+3+7+1=15}$$

▶ Both of those totals are divisible by 3, so we can reduce this fraction by a factor of 3. Divide both the numerator and denominator by 3 to get:

$$\frac{181}{1,457}$$

▶ Can we reduce that further? $1+8+1$ doesn't add up to be divisible by 3, so 3 is out. What about 9? $1+8+1$ doesn't add up to be divisible by 9 either, so 9 is out. We can try 7, but that doesn't work either. Looks like 181 is a prime number, so this is as far as we can reduce that fraction. $\frac{181}{1,457}$ is the lowest form.

Dividing Fractions

Dividing fractions is very similar to multiplying fractions. In fact, it's exactly the same except that we change the second fraction (the divisor) to its reciprocal. Multiplication and division are like reciprocal operations, so all we have to do is flip that second fraction over and then multiply the numerators straight across to get the numerator of the product and then multiply the denominators straight across to get the denominator of the product.

EXAMPLE

▶ What is $\dfrac{4}{7}$ divided by $\dfrac{2}{5}$?

▶ Flip the second fraction over and multiply.

$$\frac{4}{7} \div \frac{2}{5} = \frac{4}{7} \times \frac{5}{2}$$

$$\frac{4 \times 5}{7 \times 2} = \frac{20}{14}$$

▶ We will probably need to reduce that answer, so let's do that.

$$\frac{20}{14} \div \frac{2}{2} = \frac{10}{7}$$

▶ We may also need to convert it to a mixed number, so let's do that too.

$$\frac{10}{7} = 1\frac{3}{7}$$

BTW

Don't forget that, to divide fractions, you have to flip the second one over before you multiply!

EXERCISES

EXERCISE 2-1

Convert these improper fractions to mixed numbers.

1. $\dfrac{21}{15}$

2. $\dfrac{16}{8}$

3. $\dfrac{25}{10}$

4. $\dfrac{17}{2}$

5. $\dfrac{34}{3}$

EXERCISE 2-2

Convert these mixed numbers to improper fractions.

1. $2\dfrac{1}{5}$

2. $1\dfrac{1}{8}$

3. $5\dfrac{2}{3}$

4. $17\dfrac{1}{2}$

5. $3\dfrac{3}{5}$

EXERCISE 2-3

Fill in the blank between each pair of fractions using greater than (>), less than (<), or equals (=) symbols.

1. $\dfrac{1}{5}$ $\dfrac{2}{9}$

2. $\dfrac{3}{8}$ $\dfrac{4}{7}$

3. $\dfrac{5}{15}$ $\dfrac{4}{12}$

4. $\dfrac{5}{6}$ $\dfrac{4}{5}$

5. $\dfrac{4}{16}$ $\dfrac{3}{12}$

EXERCISE 2-4

Add or subtract these fractions as indicated. If the result is an improper fraction, convert it to a mixed number.

1. $\dfrac{1}{3} + \dfrac{1}{6}$

2. $\dfrac{2}{5} + \dfrac{1}{4}$

3. $\dfrac{2}{3} - \dfrac{1}{3}$

4. $\dfrac{3}{4} - \dfrac{2}{3}$

5. $\dfrac{4}{9} - \dfrac{1}{4}$

EXERCISE 2-5

Multiply or divide these fractions as indicated. Reduce the result to its lowest form. If the result is an improper fraction, convert it to a mixed number.

1. $\dfrac{1}{2} \times \dfrac{3}{4}$

2. $\dfrac{5}{6} \times \dfrac{4}{7}$

3. $\dfrac{1}{4} \div \dfrac{7}{8}$

4. $\dfrac{5}{8} \div \dfrac{1}{4}$

5. $\dfrac{3}{7} \div \dfrac{3}{1}$

Flashcard App

 **Decimals**

MUST ⚡ KNOW

⚡ A decimal number can show whole numbers as well as fractions of whole numbers. They allow us to be more precise in our measurements and calculations.

⚡ To compare, order, add, or subtract decimal numbers, line up the decimal points.

⚡ When multiplying or dividing decimal numbers, count the total number of places after the decimal points to determine where the decimal should go in the answer.

ou are probably already familiar with decimals from your experience with money. If an item costs $16.09, that means it costs sixteen whole dollars plus nine cents (part of a dollar). For money, the decimal point separates dollars from cents. You may also be familiar with decimals from doing long division. Decimals are similar to fractions because they can show parts as well as whole number values. Any time we divide a fraction that is less than one, the result is a decimal.

Place Value and Rounding

Decimals show numbers as powers of ten and rely on place value. The **place values** to the left of the decimal point are the ones, tens, hundreds, thousands, etc. Each place is ten times the place before it: $1 \times 10 = 10$ and $10 \times 10 = 100$ and $100 \times 10 = 1,000$. The place values to the right of the decimal point are the tenths, hundredths, thousandths, etc. Each place is the one before it divided by 10. Here is a chart showing place values.

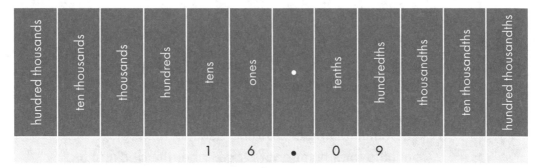

hundred thousands	ten thousands	thousands	hundreds	tens	ones	.	tenths	hundredths	thousandths	ten thousandths	hundred thousandths
				1	6	.	0	9			

We read a decimal number from left to right, using its place values. The decimal number 16.09 is read as *sixteen and nine hundredths*. In mathematics, we say *and* for the decimal point, just as we would for money when we say *sixteen dollars* **and** *nine cents*. For a decimal number that is less than one, we typically put a 0 in the ones place. Thirty-two hundredths is written as 0.32.

Since there are so many different place values, decimal numbers can have many digits. **Rounding** those longer decimal numbers is common. The rounded decimal is not quite as accurate, but it is much easier to write and to work with a rounded decimal number that has fewer digits. You should know how to round whole numbers based on their place values; for example, 1,469 rounded to the nearest hundred is 1,500.

Rounding decimals works in much the same way. Find the digit that is in the place you are rounding to and then look at the digit *after* it. If that number is less than five, round down; if that number is five or greater, round up. Rounding to tenths means the result will have only one digit after the decimal point. Rounding to hundredths means there will be two digits after the decimal point. Rounding to the thousandths means there will be three digits after the decimal point, and so on.

EXAMPLE

▶ Round 52.671 to the nearest hundredth.

▶ Find the digit in the hundredths place: 7.

▶ Look at the number after that: 1.

▶ Since one is less than five, round down: 52.671 to the nearest hundredth is 52.67.

You may be asked to round to a certain number of **decimal places** rather than to a specific place value. Rounding to one decimal place means that there will be one digit after the decimal place; it is the same as rounding to tenths. Rounding to two decimal places is the same as rounding to hundredths, and there will be two digits after the decimal point.

▶ Round 568.2968 to two decimal places.

▶ "Two decimal places" means the result will have two digits after the decimal point.

▶ Find the number that is two digits after the decimal point in the original number: 9.

▶ Look at the number after that: 6.

▶ Since 6 is in the "5 or greater" category, round up. Remember that, when we round up 9, it becomes 10, so 0.29 becomes 0.30. 568.2968 rounded to two decimal places is 568.30.

You may also be asked to round to a certain number of **significant digits** (or **significant figures**) rather than to a specific place value or number of decimal places. Significant digits are digits that are important to the measurement of the number. All nonzero digits are significant.

▶ Round 22.4691 to four significant digits.

▶ Start with the whole number part. Count four digits from left to right: 22.46.

▶ Look at the digit after the one we are rounding: 9.

▶ Since nine is in the "five or greater" category, round up: 22.4691 rounded to four significant digits is 22.47.

▶ Round 324.25 to three significant digits.

▶ "Three significant digits" means the result will have three digits in total.

▶ Start with the whole number part. Since there are already three digits in the whole number, we will be rounding to the ones place.

> ▶ Look at the digit after the one you are rounding. In this case, it will be the first digit after the decimal point, the number in the tenths place: 2.
>
> ▶ Since 2 is less than 5, round down: 324.25 rounded to three significant digits is 324.

Zeros are where significant digits get weird:

Leading 0s

- For a decimal number less than one, the 0 we put in the ones place does not count as a significant digit because it has no real measurable value. 0.34 is not different from .34—it just has a placeholder 0 by convention. There are two significant digits in 0.34.

- Zeros after a decimal point but before the first nonzero number are also not significant. These leading 0s really are just placeholders to show where the measurement starts. They aren't actually part of the measurement, so they are not considered significant. 0.0034 has only two significant digits. That may not seem to make much sense at first. Look at it this way: if we measured 0.0034 in kilometers, that's exactly the same as 3.4 meters, which clearly has only two significant digits. The 0s just show at what place value the measurement actually starts.

Trailing 0s

- Zeros at the end of a number without a decimal point are not significant. The number 2,400 has only two significant digits (2 and 4) because we do not know that the 0s represent precise measurement. The number may have been rounded to 2,400 from a larger or smaller actual number instead of containing real 0s. 2,423 rounded to the nearest hundred is 2,400, so if we see 2,400 we do not know that it is a precise measurement. In addition, 2,400 meters is the same as 2.4 kilometers, which clearly only has two significant digits.

■ Trailing 0s after a decimal point are significant. The number 13.00 has four significant digits. We know this is a precise value because if it were rounded from something else, there would be no need to put a decimal and those last two 0s on it.

■ Trailing 0s just before a decimal point are significant. The number 340. has three significant digits. The decimal point at the end tells us that 340 is a precise measurement. The number 340 without the decimal point might have been rounded, so it has only two significant digits.

Zeros Within Numbers

■ Zeros within a number, such as 104, are significant because they represent an actual value. 1.0034 has five significant digits because the 0s are in between nonzero digits and are part of the measurement.

Rounding to a certain number of significant digits means keeping track of all these fun rules!

EXAMPLE

▶ How many significant digits are in 22.460?

▶ This number ends in a 0, so we have to worry about all those rules. Do we need that 0? If not, why is it there? If it isn't important, the person who measured that value would have just said 22.46, so the 0 *is* part of the actual measurement. 22.460 has five significant digits.

▶ How many significant digits are in 0.00308?

▶ This number has a whole bunch of 0s, so we have to worry about all those rules.

▶ Do we need the 0 before the decimal point? Definitely not. We won't count that one.

▶ Do we need the two 0s right after the decimal point? Not really. They are just placeholders to tell us where the measurement starts. If 0.00308 is in kilometers, we could have written 3.08 meters instead. We don't count those either.

▶ Do we need the 0 between 3 and 8? Yes. 308 is not the same as 38. The 0 there *is* part of the measurement.

▶ Is that all the 0s? Finally? 0.00308 has three significant digits.

If you have a very long decimal value, you will almost always want to round it off in some way. Some decimals even go on forever! If we divide 1 by 3, we get 0.33333333 ..., and it just keeps repeating the three forever. A repeating decimal can be rounded off, or it can be expressed like this, with a bar over the digit or digits that repeat: $0.\overline{3}$.

Comparing and Ordering Decimals

Comparing two decimals to see which is larger or putting a group of decimals in order of their value begins with one simple step: line up the decimal points. It would be easy to incorrectly compare 2.49 and 24.6 if we did not line up the decimal points. At first glance, your eyes may see 249 and 246, and if you don't pay careful attention to the decimal point, you may think 249 is larger. However, that number is really just 2.49, or about two and a half. The other number is 24.6, or about 24 and a half. There is a big difference between the two numbers!

To compare or order decimals, follow these steps.

1. Write the numbers one on top of the other, with the decimal points lined up.
2. If you have numbers with different numbers of digits after the decimal point, remember that we can always add 0s to the end of a decimal number (after the decimal point at the very end) as placeholders so that we have the same number of places to compare. Those 0s are not significant digits, right?
3. Compare each digit from left to right, beginning with the whole numbers (if there are any).
4. When you find a difference, whichever digit is larger will be the larger number.

EXAMPLE

▶ Which is larger: 354.89 or 84.31006?

▶ Line up the decimal points.

  354.89

  84.31006

▶ If you like, we can add 0s to the end of the first number so that it has the same number of decimal places as the second number.

  354.89000

  84.31006

▶ Now it is easy to see that the first number has more place values before the decimal point. We are really comparing a number close to 355 with a number close to 84. You should have no trouble seeing that the first number is larger.

  354.89 > 84.31006

It can also be tricky to compare decimal numbers that are very close to each other, such as 1.301 and 1.302. We need to compare each number one digit at a time until we find a difference. If the whole numbers are equal, compare the tenths place. If the tenths places are equal, compare the hundredths places, then the thousandths places, and so on until you find one that is larger than the other, or there are no more places to compare. When you find a difference, you can see which is larger.

EXAMPLE

▶ Put the following in order from least to greatest:

a. 5.013   b. 5.103   c. 5.0103

▶ Line up the decimal points.

a. 5.013

b. 5.103

c. 5.0103

▶ If you like, we can add place-holder 0s to the first two numbers so that all three have the same number of place values.

a. 5.0130

b. 5.1030

c. 5.0103

▶ Compare the first digit. Each number has a 5 in the ones place. Move to the next place, the tenths place. Choice *b* has a 1 in the tenths place, while choices *a* and *c* each have a 0. That means choice *b* is the largest. Now we just need to compare choices *a* and *c*. If it helps, we can rewrite them without choice *b* in the middle.

a. 5.0130

c. 5.0103

▶ Since the ones and tenths places are equal, compare the hundredths places. Both have a 1 in the hundredths place. Compare the thousandths place. Choice *a* has a 3, and choice *c* has a 0. That means choice *a* is larger than choice *c*.

▶ Now we can put the three numbers in order. Recheck the question to see what order to use. It asks for least to greatest, so we want $c < a < b$:

$$5.0103 < 5.013 < 5.103$$

Adding and Subtracting Decimals

Addition and subtraction with decimal numbers both begin with the simple step of lining up the decimal points. As with comparing, we can add placeholder zeros after the decimal point at the very end of a number so that both numbers have the same number of decimal places. Then add the numbers just as we do with whole numbers. When we get to the decimal point, we just bring it down and put it in the result in the same place it was in the original numbers.

EXAMPLE

▶ Noah wants to add up the expenses for his new car this month. His monthly car payment is $328.64, and his monthly insurance payment is $76.42. If he spent $117 on gas, what is the total he spent on the car this month?

▶ To find the total, we will need to add the three values for the car payment, insurance, and gas. Write those values as an addition problem and be sure to line up the decimal points. If you like, we can add a decimal point and two 0s to the end of the gas value so that it has the same number of decimal places as the other values.

$$328.64$$
$$76.42$$
$$+ \ 117.00$$

▶ Now we can add as we normally would with whole numbers. Just remember to keep the decimal point in the same position in the result; in other words, there should be two digits after the decimal point.

$$328.64$$
$$76.42$$
$$+ \ 117.00$$
$$\overline{522.06}$$

Subtraction works the same way. Line up the decimal points. If you like, add placeholder 0s as needed so that all numbers have the same number of decimal places. Then subtract the numbers just as we would with whole numbers. Keep the decimal point in the same position in the result.

BTW

Remember to line up the decimal points before you compare, order, add, or subtract decimal numbers.

EXAMPLE

▶ Isabella received $100 for her birthday from her grandparents. She spent $64.91 on some new clothes. How much money does she have left?

▶ To find out how much she has left, we will need to subtract what she spent from the original $100. Write those values as a subtraction problem and be sure to line up the decimal points. If you like, we can add a decimal point and two 0s to the end of $100 so that it has the same number of decimal places as the other value.

100.00
−64.91

▶ Now we can subtract as we normally would with whole numbers. Just remember to keep the decimal point in the same position in the result; in other words, there should be two digits after the decimal point.

100.00
−64.91
35.09

If you only need an approximate answer, don't forget that you can estimate! Round the decimals to the nearest whole number before you add or subtract.

EXAMPLE

▶ Kim buys a candy bar that costs $0.89 and a bottle of water that costs $1.89. She gives the cashier $5.00. What will her change be?

a. $0.89

b. $1.49

c. $2.22

d. $3.78

▶ The candy bar can be rounded up to $1, and the water can be rounded up to $2. That's a total of $3. If she paid with a $5, she will get $2 back. Choose choice c.

▶ Since we rounded both prices slightly up, the real total will be slightly less than $3, so she will get slightly more than $2 back. Choice c is the only one that is possible.

Multiplying Decimals

Multiplication with decimal numbers is not that much more complicated than multiplication with whole numbers. In fact, it is exactly the same except for the additional step of figuring out where to put the decimal point in the product. We do *not* have to line up the decimal points as we do for addition and subtraction.

The technique for placing the decimal point in the product is to count how many decimal places there are after the decimal points in the factors (the numbers we are multiplying together). For example, the number 12.54 has two decimal places and the number 54.1 has one decimal place. That is a total of three decimal places. When we find the product, we will start from the right and count three places toward the left. Place the decimal point *after* the first three places. Remember to count from right to left.

EXAMPLE

▶ What is 12.54 multiplied by 54.1?

▶ Set up the multiplication as we would with whole numbers. Put the number with the most digits on top to make the multiplication easier. Do not line up the decimal points.

$$\begin{array}{r} 12.54 \\ \times\ \ 54.1 \\ \hline \end{array}$$

▶ Now multiply as we normally would with whole numbers.

$$
\begin{array}{r}
12.54 \\
\times\ 54.1 \\
\hline
1254 \\
50160 \\
627000 \\
\hline
678414 \\
\end{array}
$$

▶ Finally, add up the total number of decimal places there were in the factors. 12.54 has two decimal places, and 54.1 has one decimal place, so that is a total of three decimal places. In the product, count three places from the right and place the decimal point *after* that third digit.

$$
\begin{array}{r}
12.54 \\
\times\ 54.1 \\
\hline
678.414 \\
\end{array}
$$

When we multiply an amount of money by some number, we will also need to round off the product to two decimal places, since money is always expressed that way.

EXAMPLE

▶ At a school fun run for charity, Ms. Furse pledged to donate $.75 for each mile that her 7th grade class ran. Her students ran a total of 68.25 miles. How much will Ms. Furse donate?

▶ Set up the multiplication as we would with whole numbers. Put the number with the most digits on top to make the multiplication easier. Do not line up the decimal points.

$$68.25$$
$$\times \ 0.75$$

▶ Now multiply as we normally would with whole numbers.

$$68.25$$
$$\times \ 0.75$$
$$34125$$
$$\underline{477750}$$
$$511875$$

▶ Add up the total number of decimal places in the factors: 4. In the product, count four places from the right and place the decimal point *after* that fourth digit.

$$68.25$$
$$\times \ 0.75$$
$$\overline{51.1875}$$

▶ Since money is expressed in dollars and cents, we need to round the product to two decimal places: $51.19.

Multiplying a Decimal Number by a Power of 10

Place value is all based on powers of 10. The ones place times 10 is the hundreds place. The hundreds place times 10 is the thousands place, and so on. The same is true for the place values after the decimal point. Tenths divided by 10 are hundredths. Hundredths divided by 10 are thousandths, and so on.

How does this knowledge help us? It can give us a shortcut when we are asked to multiply a decimal number by a power of 10. When we multiply a decimal number by 10, we are really just moving the decimal point one space to the right. Try it: $0.4 \times 10 = 4$. For each power of 10, we move one place to the right in the decimal number. When we multiply a decimal number by 100, we can move the decimal point two spaces to the right because 100 is two powers of 10. We can also just count the 0s and move that many places to the right. $1,000 =$ three 0s $=$ move three places to the right.

$$93.685 \times 1,000$$

Since 1,000 is three powers of 10 (and has three 0s), move the decimal in 93.685 three places to the right.

$$93.685 \times 1,000 = 93,685$$

The same is true for decimal powers of 10, but we move the decimal point to the *left* rather than to the right. Count the number of decimal places in the number involving the power of 10 and move to the left the same number of places in the decimal number: 0.01 has two decimal places, so we would move the decimal point in the other number two places to the left.

$$24.56 \times 0.001$$

Since 0.001 has three decimal places, move the decimal in 24.56 three places to the left.

$$24.56 \times 0.001 = 0.02456$$

If you only need an approximate answer, don't forget that you can estimate! Round a decimal number to its highest place value.

EXAMPLE

▶ What is $411.86 \times .224$?

a. 1.84

b. 92.26

c. 823.71

d. 92,256.64

▶ Since the answer choices are far apart from one another, we can estimate. Round each number to its highest place value:

$$411.86 = 400$$

$$0.224 = 0.2$$

▶ Now multiply: $400 \times 0.2 = 80$. The only answer choice even close to that is choice *b*.

Dividing Decimals

Division with decimal numbers is similar to division with whole numbers except for the step of figuring out where to put the decimal point. For multiplication, we did that after we found the product. For division, we must do it *before* we divide.

Dividing a Decimal by a Whole Number

When we divide a decimal number by a whole number, only the **dividend** (the number we are dividing) has decimal places. Place the decimal point for the **quotient** (the answer) directly above the decimal point in the dividend. That's all we have to do. Then we can divide as we normally would:

EXAMPLE

▶ $6 \overline{)241.8}$

▶ Divide as you normally would.

```
        40.3
   6 ) 241.8
       24
       ‾‾
       01
        0
       ‾‾
       18
       18
       ‾‾
        0
```

Dividing a Whole Number by a Decimal

When we divide a whole number by a decimal number, only the **divisor** (the number we are dividing by) has decimal places, so we only need to count those decimal places. Move the decimal point all the way to the right of the divisor, turning it into a whole number. Add 0s to the dividend until there are as many 0s as decimal places in the divisor. For example, the divisor 0.89 has two decimal places, so we move the decimal two places to the right and add two 0s to the dividend.

EXAMPLE

▶ 　 0.8) 82

▶ Since the divisor has one decimal place, you will move that decimal point one place to the right and add one 0 to the dividend.

　 8) 820

▶ Now divide as you normally would.

```
        102.5
   8 ) 820.0
       8
       ‾‾
       02
        0
       ‾‾
       20
       16
       ‾‾
        40
        40
        ‾‾
         0
```

Dividing a Decimal by a Decimal

When we divide a decimal number by another decimal number, they both have decimal places to keep track of. Move the decimal point in the divisor all the way to the right, turning it into a whole number. Move the decimal point in the dividend to the right the same number of places you did for the divisor. Put the decimal point for the quotient directly above the new decimal point in the dividend.

▶ $0.6\overline{)25.845}$

▶ There is one decimal place in the divisor, so move it to the right one place. Move the decimal point in the dividend one place to the right too. Now put a decimal point for the quotient directly above the new decimal point in the dividend.

$$6\overset{\Large .}{\overline{)258.45}}$$

▶ Divide as you normally would.

```
          43.075
    6) 258.450
       24
        18
        18
         04
          0
          45
          42
           30
           30
            0
```

If you only need an approximate answer, don't forget that you can estimate! Round a decimal number to its highest place value.

▶ What is 485.23 divided by 4.8?

a. 1.01

b. 10.11

c. 101.09

d. 1010.90

▶ Since the answer choices are far apart from one another, we can estimate. Round each number to its highest place value:

$485.23 \rightarrow 500$

$4.8 \rightarrow 5$

▶ Now divide: $500 \div 5 = 100$. The only answer choice even close to that is choice *c*.

Dividing a Decimal Number by a Power of 10

Just as with multiplication, understanding place value can give us a shortcut when we are asked to divide a decimal number by a power of 10. The only difference is the direction we move the decimal point. Division is the opposite of multiplication. When we divide, we move the decimal in the opposite direction from the way we move it in multiplication. When we divide a decimal number by 10, we are really just moving the decimal point one space to the left. Try it: $23.6 \div 10 = 2.36$. For each power of 10, move one place to the left in the decimal number. When we divide a decimal number by 100, we can move the decimal point two spaces to the left because 100 is two powers of 10. We can also just count the 0s and move that many places to the left: 1,000 → three 0s → move three places to the left.

$$8,736.25 \div 1,000$$

Since 1,000 is three powers of 10 (and has three 0s), move the decimal in 8,736.25 three places to the left:

$$8,736.25 \div 1,000 = 8.73625$$

The same is true for decimal powers of 10, but we will move the decimal point to the *right* rather than to the left. Count the number of decimal places in the power of 10 number and move to the right the same number of places in the decimal number. 0.01 has two decimal places, so we would move the decimal point in the other number two places to the right:

$$5.6 \div 0.01$$

Since 0.01 has two decimal places, move the decimal in 5.6 two places to the right:

$$5.6 \div 0.01 = 560$$

Converting Between Decimals and Fractions

At the beginning of the chapter, I said that decimals are similar to fractions because they can show parts of numbers as well as whole numbers. In fact, a decimal can show exactly the same number as a fraction, just in a different way. It is helpful to know how to change a fraction to a decimal and how to change a decimal into a fraction. That flexibility allows you to decide which will be easier to use in your calculations.

To change a fraction to a decimal, just use the fraction bar as a division sign. The fraction $\frac{3}{4}$ can be read as three divided by four. Set up your division problem as 3 divided by 4, and the quotient you find will be the decimal value for $\frac{3}{4}$:

$$\begin{array}{r} 0.75 \\ 4\overline{)3.00} \\ \underline{28} \\ 20 \\ \underline{20} \\ 0 \end{array}$$

$\frac{3}{4}$, then, is equal to 0.75.

▶ Change $\frac{5}{8}$ to a decimal.

▶ Set up the division 5 divided by 8.

$$\begin{array}{r} 0.625 \\ 8\overline{)5.000} \\ \underline{48} \\ 20 \\ \underline{16} \\ 40 \\ \underline{40} \\ 0 \end{array}$$

▶ $\frac{5}{8}$, then, is equal to 0.625.

If the fraction has a power of 10 in the denominator, this is even easier. Remember that dividing a number by ten is the same as moving the decimal point one place to the left? We can use that here as well. If we want to change $\frac{3}{10}$ to a decimal, all we are doing is dividing 3 by 10, which means moving the decimal point one place to the left: 3 becomes 0.3.

▶ Change $\frac{658}{100}$ to a decimal.

▶ Dividing a number by 100 is the same as moving the decimal point two places to the left. 658 becomes 6.58.

▶ Change $\dfrac{5}{1000}$ to a decimal.

▶ Dividing a number by 1,000 is the same as moving the decimal point three places to the left. We will need to add some leading 0s here since 5 doesn't have three decimals. 5 then becomes 0.005.

Changing a decimal to a fraction depends on place value, so we have to pay attention. The first place after a decimal point is the tenths place, which means $\dfrac{1}{10}$. A decimal value of 0.7 has a 7 in the tenths place, so that means $\dfrac{7}{10}$. A decimal value of 0.75 has two decimal places (tenths and hundredths), so we need to show that as a fraction with 100 as the denominator: $0.75 = \dfrac{75}{100}$. I like to use a shortcut here. I just count the number of decimal places I have and use the same number of 0s after the 1 on the bottom of my fraction.

EXAMPLE

▶ Change 0.865 to a fraction.

▶ Count the number of decimal places. 0.865 ends with the thousandths place, so we will need to show that as a fraction with 1,000 as the denominator. In other words, since there are three decimal places, the number on the bottom of the fraction should be a 1 with three 0s.

$$\dfrac{865}{1,000}$$

▶ This fraction can be reduced, so let's do that. Both the numerator and denominator are divisible by 5.

$$\dfrac{865}{1,000} = \dfrac{173}{200}$$

EXERCISES

EXERCISE 3-1

Which place is the digit 3 in for each of the following numbers?

1. 3.468

2. 0.317

3. 1.8673

4. 13,067.84

5. 2.63

EXERCISE 3-2

Put each set of numbers in order from greatest to least.

1. 7.021	72.1	7.027
2. 0.967	0.9067	0.0967
3. 1.293	1.2904	1.2906
4. 0.0846	0.08	0.084
5. 263.6	2.636	263.16

EXERCISE 3-3

Add or subtract as indicated. Round the answer to two decimal places.

1. $34.56 + 23.98$

2. $575.234 + 1,290.35$

3. $2.3 - 2.1$

4. 0.93485 − 0.02348

5. 6.23 − 0.023

EXERCISE 3-4

Multiply as indicated.

1. 23.3 × 0.02

2. 568 × 0.1

3. 13.5 × 1.9

4. 654.89 × 1.7

5. 84.6 × 100

EXERCISE 3-5

Divide as indicated. Round the answer to three significant digits.

1. 52.6 ÷ 4

2. 8 ÷ 0.4

3. 23.6 ÷ 1.6

4. 98.7 ÷ 0.1

5. 762 ÷ 100

EXERCISE 3-6

Convert each of these fractions to decimals. Round the answer to the nearest tenth.

1. $\dfrac{1}{8}$

2. $\dfrac{17}{18}$

3. $\dfrac{6}{10}$

4. $\dfrac{5}{16}$

5. $\dfrac{34}{100}$

EXERCISE 3-7

Convert each of these decimals to fractions.

1. 0.8

2. 1.2

3. 345.23

4. 8.2345

5. 105.00

 **Percents**

MUST ⚡ KNOW

⚡ A percent is similar to a fraction or a decimal because it shows us part of a whole. For percents, however, the whole is 100: "20%" means 20 out of 100.

⚡ We can convert a percent into a fraction by dividing the percent by 100. We can convert a percent into a decimal number by moving the decimal point two places to the left.

⚡ We use percents often in our everyday lives to calculate interest, discounts, tax, and tips.

Percent means *for every 100*. If you have ever heard a meteorologist say that there is a 40% chance of rain today or have been in a store having a 25% off sale, you are aware of percents. Percents are similar to decimals and fractions because they can show parts as well as whole number values. A fraction shows part of a whole of 1, while a percent shows part of a whole of 100. "80%" means 80 out of 100. You can also express that as a fraction, $\frac{80}{100}$, or as a decimal, 0.80.

We will cover how to convert among fractions, decimals, and percents in just a moment. After that, we will discuss how to calculate a certain percent of a number, how to find the original number from which a percent was calculated, and how to find what amount is a certain percent of a number. We will also discuss how to calculate interest; how to figure out discounts, tax, and tips; and how to increase or decrease an amount by a certain percent.

Converting Between Percents and Decimals

Percents can easily be expressed as decimals. You know how to convert from a fraction to a decimal, so if you know $20\% = \frac{20}{100}$, then you can divide 20 by 100 to get 0.20. If you want to find 5% as a decimal, remember that $5\% = \frac{5}{100}$, and then you can divide 5 by 100 to get 0.05. If you want to find 12.5% as a decimal, remember that $12.5\% = \frac{12.5}{100}$, and then you can divide 12.5 by 100 to get 0.125. Are you starting to see a pattern here? Each time, since we are dividing by 100, we are moving the decimal point from the percent value two places to the left. That can give us a shortcut. Any percent can be turned into a decimal by moving the decimal point two places to the left:

$$45\% = 0.45$$
$$89\% = 0.89$$
$$2\% = 0.02$$
$$112\% = 1.12$$

EXAMPLES

▶ Express 1.8% as a decimal number.

▶ Remember that 1.8% is the same as $\dfrac{1.8}{100}$. When we divide 1.8 by 100, we are moving the decimal point two places to the left.

$$1.8\% = 0.018$$

▶ Let's look at another example. Express 250% as a decimal number.

▶ Remember that 250% is the same as $\dfrac{250}{100}$. When we divide 250 by 100, we are moving the decimal point two places to the left.

$$250\% = 2.5$$

To convert a decimal number to a percent, we just reverse the process and move the decimal point two places to the right.

EXAMPLES

▶ Express 0.39 as a percent.

▶ All we have to do here is move the decimal point two places to the right and add the percent symbol.

$$0.39 = 39\%$$

▶ Here's another example. Express 0.538 as a percent.

▶ When we move the decimal point two places to the right and add the percent symbol, we get 53.8%. It is pretty common to round the decimal off to the hundredths place (because percent is "per hundred"), so let's do that with this one.

$$0.538 = 0.54 = 54\%.$$

Converting Between Percents and Fractions

The most important thing to remember is that a percent is out of a whole of 100. Any percent can be written as a fraction with 100 as the denominator. 100% can be written as $\frac{100}{100}$, which equals one. 65% can be written as $\frac{65}{100}$, and 5% can be written as $\frac{5}{100}$. We can even have percents greater than one whole: 120% can be written as $\frac{120}{100}$.

EXAMPLE

▶ Express 46% as a fraction.

▶ To convert a percent to a fraction, write the percent over 100.

$$46\% = \frac{46}{100}$$

▶ This fraction can be reduced, so let's do that: $\frac{46}{100} = \frac{23}{50}$.

Let's try a more difficult one that involves decimals.

EXAMPLE

▶ Express 5.34% as a fraction.

▶ To convert a percent to a fraction, write the percent over 100.

$$5.34\% = \frac{5.34}{100}$$

▶ We do not want to have a decimal number as part of a fraction, so what can we do to change it? To increase 5.34 to 534, we would need to multiply by 100, so let's multiply this fraction by one in the form of $\frac{100}{100}$.

$$\frac{5.34}{100} \times \frac{100}{100} = \frac{534}{10,000}$$

▶ This fraction can be reduced, so let's do that.

$$\frac{534}{10,000} = \frac{267}{5,000}$$

To change a fraction into a percent, the fraction must have a denominator of 100. If you have $\frac{13}{100}$, that is the same as 13%. $\frac{27}{100}$ equals 27%. What if you have a fraction that does not have 100 in the denominator? You may be able to change it into a fraction that does.

EXAMPLE

▶ Convert $\frac{1}{5}$ to a percent.

▶ Since $\frac{1}{5}$ does not have 100 in the denominator, we need to change it into a form that does have 100 in the denominator. Remember that we can multiply anything by one without changing the original number, so we want to multiply $\frac{1}{5}$ by a fraction that equals one in order to give us 100 in the denominator.

▶ How can we do that? Currently, the denominator is five. Five times what number gives us 100? Twenty does. Let's multiply $\frac{1}{5}$ by $\frac{20}{20}$ then.

$$\frac{1}{5} \times \frac{20}{20} = \frac{20}{100}$$

▶ Now that the fraction has 100 in the denominator, we can say that $\frac{1}{5} = 20\%$.

What about a fraction that can't be multiplied to make 100 in the denominator? There is a solution for that, too.

▶ Convert $\frac{1}{3}$ to a percent.

▶ Since $\frac{1}{3}$ does not have 100 in the denominator, we need to change it somehow to find the percent. Can we multiply $\frac{1}{3}$ by a fraction that equals one in order to give us 100 in the denominator? No, because the denominator is 3. Three times what number gives us 100? That number would not be a whole number, so that doesn't work. When that happens, just convert the fraction into a decimal first, and then convert the decimal to a percent.

$$\frac{1}{3} = 1 \div 3 = 0.333...$$

▶ We can convert the decimal into a percent by moving the decimal point two places to the right. Since this is a repeating decimal, we can either say $0.\overline{3} = 33.\overline{3}\%$, or we can round the repeating decimal to the nearest hundredth (because percent is "per hundred").

$$0.333... = 0.33 = 33\%$$

When we work with percents, being able to convert the percent either to a fraction or to a decimal allows us to choose which way will make the calculations easier. Flexibility is good.

Some fraction/decimal/percent conversions are very common, and it may benefit you to memorize them so that you can convert from one form to another very quickly. Here is a chart of the most common ones:

Fraction	Decimal	Percent
$\dfrac{1}{5}$	0.2	20%
$\dfrac{1}{4}$	0.25	25%
$\dfrac{1}{3}$	$0.\overline{3}$	33%
$\dfrac{1}{2}$	0.5	50%
$\dfrac{2}{3}$	$0.\overline{6}$	67%
$\dfrac{3}{4}$	0.75	75%

Calculating Percent

Now that you know a percent can be written as a fraction or as a decimal, let's learn how to find a certain percent of a number. You may see a question that asks you to find 20% of 360.

We can write 20% either as $\dfrac{20}{100}$ or as 0.20. What does "of 360" mean? If I asked you to find $\dfrac{1}{2}$ of 10, what would you do? You would divide ten by two, which is the same as multiplying 10 by $\dfrac{1}{2}$. Think of the word "of" like the word "multiply": "$\dfrac{1}{2}$ of 10" means $\dfrac{1}{2}$ multiplied by 10, or $\dfrac{1}{2} \times 10 = 5$. The phrase "of 360" means the same as "multiply by 360." To find 20% of 360, you can multiply 360 by $\dfrac{20}{100}$ or by 0.20, whichever is easier for you to do:

$$\dfrac{360}{1} \times \dfrac{20}{100} = \dfrac{7,200}{100} = 72$$

> **BTW**
>
> Finding 10% of a number is simple since 10% uses a power of 10. Just move the decimal point one place to the left: 10% of 16 is 1.6, and 10% of 232 is 23.2.

Or, if you would rather calculate using the percent as a decimal number:

$$
\begin{array}{r}
\overset{1}{360} \\
\times \quad .2 \\
\hline
\cancel{720}
\end{array}
$$

There is one digit after the decimal point, so the same should be true of the product: 72.0, or just 72.

Let's try a couple more to be sure you are comfortable finding a certain percent of a number.

▶ In Mr. Webster's art class, 12% of the 40 students were selected to show their paintings at a local art show. How many students were selected?

▶ First, let's make sure we understand what the question is asking. We need to find 12% of 40. We can do this as a fraction and write $\dfrac{12}{100} \times \dfrac{40}{1}$.

▶ Remember the Commutative Property of Multiplication? It says that we can multiply numbers in any order. That means $\dfrac{12}{100} \times \dfrac{40}{1}$ can be written as $\dfrac{12 \times 40}{100 \times 1}$ or as $\dfrac{40 \times 12}{100 \times 1}$. Then we can say:

$\dfrac{40}{100} \times \dfrac{12}{1} = \dfrac{4}{10} \times \dfrac{12}{1} = \dfrac{48}{10} = 4.8$. Since we cannot select 4.8 students, we can round the decimal number to 5.

▶ All that is a complicated way of saying that when we multiply, we can reduce not just top to bottom but also crossways.

$$\frac{12}{10\cancel{0}} \times \frac{4\cancel{0}}{1} = \frac{48}{10} = 4.8$$

▶ You may prefer to do this one as a decimal. If so, multiply .12 by 40.

$$\begin{array}{r} .12 \\ \times\, 40 \\ \hline 4.80 \end{array}$$

Since there are two digits after the decimal point, put the decimal point in the product in the same spot: 4.80, or just 4.8.

▶ Let's try another problem. What is 5% of 20% of 200?

▶ That may look tough, but don't worry. Write it out as a fraction calculation. Remember that "of" means multiply.

$$\frac{5}{100} \times \frac{20}{100} \times \frac{200}{1}$$

▶ Now reduce. Remember that we can reduce crossways too for multiplication.

$$\frac{5}{10\cancel{0}} \times \frac{2\cancel{0}}{10\cancel{0}} \times \frac{2\cancel{00}}{1} = \frac{5 \times 2 \times 2}{10} = \frac{20}{10} = 2$$

▶ See? That wasn't tough at all!

 IRL Grades on tests and assignments in school are usually calculated based on percents. On a test with 100 questions, if you get 87 of the answers correct, you got 87% of them correct and your score is 87.

What happens when there are not 100 questions on the test? Let's say there are 40 questions on the test, and you miss four of them. How is your grade calculated? Usually, the teacher will find out how much each question is worth by dividing 100 by the number of questions.

For a 40-question test, the teacher would divide 100 by 40 to find that the questions are worth 2.5 points each. If you missed 4, that is 4 × 2.5 = 10. Subtract 10 from 100, and you get a score of 90. You answered 90% of the 40 questions correctly.

Interest

Interest is basically a fee paid for borrowing. When we borrow money from a bank or use a credit card, the bank will charge interest. The interest is what the bank earns in return for allowing us to borrow its money. The amount we borrow is called the **principal**, and the interest is calculated by percent. For example, a $1,000 loan may have a 2% interest rate. That means that the borrower must pay back the $1,000, plus another 2% of the $1,000.

When we lend money, it works the same way, but the interest is paid to us rather than to the bank since *we* are the lenders. The interest is what we earn in return for allowing someone else to use our money.

When we put money in a savings account or even in some other types of bank accounts, we may earn interest. We earn interest because we are allowing the bank to use our money temporarily to lend to others.

There are several types of interest calculations, such as simple interest, compound interest, and annual percentage yield. The only one you will be asked to calculate at this level is **simple interest**. Simple interest is a fee based on a certain percent of the **principal** (the original amount) paid yearly. To calculate simple interest (I), we multiply the principal (p) by the interest rate (r, the percent), times the amount of time in years (t). The formula for simple interest is:

$$I = p \times r \times t$$

> **BTW**
>
> *Don't forget that, when calculating interest, time is expressed in years. If the time is six months, don't use 6 for t. Six months is $\frac{1}{2}$ a year, so use 0.5 for t.*

EXAMPLE

▶ Laura takes out a loan of $500. The bank charges 4% simple interest yearly on the loan. If Laura pays the loan back in one year, how much does she pay in interest?

▶ All we have to do here is calculate the amount of interest, 4% of $500.

$$I = p \times r \times t$$

$$I = \frac{500}{1} \times \frac{4}{100} \times 1 = \frac{4 \times 5}{1} = 20$$

▸ Laura pays $20 interest on the $500 loan.

To find the total amount paid on a loan, including the interest, find the amount of interest and add it to the principal.

EXAMPLE

▸ Jim opens a new savings account with $120. The bank pays 15% simple interest per year. If Jim does not add any more money and does not withdraw any money, what is the total amount, including interest, he will have in the account at the end of the year?

▸ First, calculate the amount of interest, 15% of $120. That one may be easier to do as a decimal.

$$I = p \times r \times t$$

$$I = 120 \times 0.15 \times 1 = 18$$

▸ Now add the $18 interest to the $120 principal.

$$18 + 120 = 138$$

▸ Jim will have $138 in his account at the end of the year.

We can also find interest for periods longer or shorter than one year. All we need to do is make sure that the amount of time we use for the formula is expressed in years. In the example we just did, if the time was three years, we would just multiply the interest rate times the principal, times 3:

$$I = 120 \times 0.15 \times 3 = 54$$

Jim would earn $54 in interest and have a total of $174 in his account.

▶ Jack borrows $400 for a period of six months at a rate of 6.5%. What is the total amount he pays back?

▶ First, let's make sure we know which numbers to use. The principal is 400. The interest rate is 6.5%, so let's do that as a decimal: $r = 0.065$. The time is six months, but we need to express that in years, so $t = 0.5$ years.

▶ Calculate the amount of interest.

$$I = p \times r \times t$$

$$I = 400 \times 0.065 \times 0.5 = 13$$

▶ Now add the $13 interest to the $400 principal.

$$13 + 400 = 413$$

▶ Jack will pay back $413.

Discounts, Taxes, and Tips

As you can see, being able to calculate the amount of interest you will have to pay is a useful thing to be able to do, especially if you are on a tight budget. The same is true with shopping. You may want to find the price of an item that has been discounted by a certain percent. Then you will need to be able to calculate the amount of sales tax you will be charged. If you eat at a restaurant, you will need to know how much to tip the server. Discounts, taxes, and tips are usually based on percents, so all these are practical, everyday applications.

Discounts

A **discount** is a deduction from the cost of something. If you are shopping and see a sign that says "Everything 25% Off," what does that actually mean? It means that the original price of an item is discounted by an amount equal to 25% of the original price. To calculate the amount of that discount, we need to multiply the original price of the item by 25%. To find the total amount the item now costs, we subtract the discount from the original price.

EXAMPLE

▶ Elena wants to buy a pair of sunglasses. She finds a pair she likes in the clearance section of the store, where everything is discounted by 30%. The original price of the sunglasses was $16. What is the discounted price?

▶ First, let's make sure we understand what the question is asking. We need to find 30% of $16 and then subtract that discount from $16 to find the new price.

▶ Calculate the discount. Find 30% of $16. We can do that either as a fraction or as a decimal.

$$\frac{30}{100} \times \frac{16}{1} = \frac{3}{10} \times \frac{16}{1} = \frac{48}{10} = 4.8$$

▶ Or $0.3 \times 16 = 4.8$.

▶ The discount is $4.80. Now subtract the amount of the discount from the original $16 price.

$$\$16.00 - \$4.80 = \$11.20$$

▶ The discounted price is $11.20.

▶ Another way to do discounts is a bit faster. Remember that percents are out of 100? A discount of 30% is like 100% − 30%, which is 70%. A discount of 30% from the original 100% means we are paying only 70% of the original amount. To find the discounted price, we could just find 70% of $16.

$$0.7 \times 16 = 11.20$$

▶ The discounted price is $11.20.

▶ No subtraction necessary—fewer steps always makes me happy! This is the way I calculate discounts when I am shopping.

Tax

A **tax** is an additional fee that you pay when you buy things or earn income. When you have a job, you will probably have to pay income tax to the government. For most consumer goods that you buy (other than food), you have to pay a sales tax. Many services, such as entertainment and restaurants, also charge taxes. Sales tax is a fee that is in addition to the marked price of the item. State governments decide how much will be charged. Local governments may also add sales tax.

When you buy a book that is marked $19.99, that's the before-tax price. How do you figure out what the actual price is before you get to the register? When you have a job and you earn $40,000 a year, how much income tax will you have to pay? To figure these things out, we will need to know the tax rate, which is expressed as a percent.

EXAMPLE

▶ Erlene has $20. She wants to buy a shirt that is marked $18, but she knows there will be an 8% tax added. Will she have enough money to buy the shirt?

▶ Let's calculate the amount of tax that will be added. Find 8% of $18. It may be easier to do as a decimal than as a fraction. Since there are 2 digits after the decimal point, put the decimal point in the product in the same spot: 1.44.

$$
\begin{array}{r}
\overset{6}{} \\
18 \\
\times\ .08 \\
\hline
1.44
\end{array}
$$

▶ The tax will be $1.44. Add that to the price of the shirt.

$$\$18.00 + \$1.44 = \$19.44$$

▶ The total cost will be $19.44. Erlene does have enough money.

▶ We can find the total amount with only one calculation if we multiply $18 by 1 (the cost of the item) plus the tax rate. Multiply 18 by 1.08. Since there are 2 digits after the decimal point, put the decimal point in the product in the same spot: 19.44.

$$
\begin{array}{r}
\overset{6}{} \\
18 \\
\times\ 1.08 \\
\hline
144 \\
1800 \\
\hline
19.44
\end{array}
$$

▶ If we do this, we only have one calculation to find the total price including the tax, rather than finding the tax and then adding it to the original price. Fewer steps = happiness!

To find the amount of tax you will pay on your income, you need to know the tax rate. The federal government requires most people to pay income tax, and some states also have a state income tax. Those tax rates vary according to the amount of income earned.

EXAMPLE

▶ Elba Dean earns $57,000 per year as a personal trainer. If she pays 11% income tax, approximately how much tax will she owe?

a. $570

b. $627

c. $6,200

d. $11,400

▶ To find the exact amount of tax she owes, we would need to find 11% of $57,000. However, this question asks for an *approximate* amount.

▶ Finding 10% is easy, so let's do that. Simply move the decimal point over one place to the left: $5,700. The only choice close to that is choice *c*.

 IRL When you are out shopping and need to know how much something will cost after the sales tax is added, it may be much easier to simply estimate the tax.

You will need to know the sales tax rate in your area. Let's say you pay state and local tax for a combined rate of 8.5%. You can estimate the tax at 10% for an easy calculation. Just move the decimal point in the price of the item to the left by one place, and you have 10%. Ten percent of a $12.00 item is $1.20, so the total price will be a bit less than $13.20.

By rounding up from 8.5% to 10%, we also ensure that we are estimating a little higher than the actual tax, so we will definitely know whether we have enough money to purchase the item.

Tips

A **tip** is a bonus added to the cost of a service to reward good performance by the staff of the business. When we eat out at a sit-down restaurant, it is customary to tip the server. It is also common to tip hotel staff, salon workers, parking attendants, and other service providers. The most common situation is probably tipping at restaurants. A standard tip is 15%, though many people tip 20% for good service.

EXAMPLE

▶ Bob takes his family out to dinner, and the bill is for $112. If Bob wants to leave a 15% tip for the server, how much will his total be?

▶ You have some options here. You can find 15% of $112 and then add it to the total. You can find 115% of the $112 and get the total amount directly. You can do either of those things as fractions or as decimals. None of that sounds easy unless you have a calculator handy.

▶ Here is another option: You can break up the 15% into 10% plus 5% for a much easier calculation that you can probably do in your head. It will really impress the people you are having dinner with!

▶ 10% of $112 is $11.20. 5% will be half of that, so $11.20 divided by 2 equals $5.60. Add the 10% and 5% together: $11.20 +5.60 = $16.80, for the tip. Add that to the bill to get: $112.00 + $16.80 = $128.80.

▶ If you have enjoyed good service—or you enjoy easier math—tipping 20% is even easier than tipping 15%. Find 10% of the total and then double it.

▶ 10% of $112 is $11.20. Double it to get $22.40, for a 20% tip. Add that to the bill to get: $112.00 + $22.40 = $134.40.

▶ We can always round the number too for convenience. We could pay $134.00 or $135.00. There is no rule that we have to tip *exactly* 20%.

Percent Increase and Decrease

There is one more thing we need to be able to do with percents: calculate percent change. If a value goes up or down over time, we may want to know by what percent the value increases or decreases.

Let's say you bought a house for $500,000 and sold it a few years later for $600,000. To find the change in the price, you can just subtract: $600,000 − $500,000 = $100,000. By what percent did the value of your house increase? Any increase or decrease is going to be in terms of the original amount, so to figure out the percent change, we take the difference between the original value and the current value and then divide it by the original value. This will result in a decimal number, so we will also need to convert it to a percent:

$$\text{percent change} = \frac{\text{difference}}{\text{original}}$$

EXAMPLE

> Let's finish the example we started. You bought a house for $500,000 and sold it a few years later for $600,000. By what percent did the value of your house increase?
>
> Use the formula just shown. First, find the difference in the prices.
>
> $$600,000 - 500,000 = 100,000$$
>
> Now divide that by the original price.
>
> $$\frac{100,000}{500,000} = \frac{100}{500} = \frac{1}{5} = 0.2 = 20\%$$
>
> The value of your house increased by 20%. Nice!

What if it isn't so clear which value is the "original" value? You can use the wording of the question to help you figure out which of the two numbers to use for the original. If the question asks for percent *increase*, then the original value must be the smaller one in order for it to have increased. If the question asks for percent *decrease*, then the original value must be the larger one in order for it to have decreased.

EXAMPLE

▶ The number of students at Smithville High School changed to 2,200 from 2,300. By what percent did the number of students decrease?

▶ It would be easy to assume that 2,200 is the original value because it is listed first, but let's think about this carefully. The question asks for a percent *decrease*, so the original value must have been larger in order to decrease. Plus, if the number changed *from* 2,300 to 2,200, then 2,300 must be the original.

▶ Use the percent change formula. First, find the difference in the two values.

$$2{,}300 - 2{,}200 = 100$$

▶ Now divide that by the original value. Round off the decimal quotient to the hundredths place.

$$\frac{100}{2{,}300} = \frac{1}{23} = 0.04 = 4\%$$

EXERCISES

EXERCISE 4-1

Convert each of the following percents to a decimal.

1. 30%

2. 16%

3. 2.5%

4. 120%

5. 0.5%

EXERCISE 4-2

Convert each of the following decimals to a percent.

1. 0.38

2. 0.539

3. 1.246

4. 0.06

EXERCISE 4-3

Convert each of the following percents to a fraction.

1. 30%

2. 2.5%

3. 120%

4. 0.5%

EXERCISE 4-4

Convert each of the following fractions to a percent.

1. $\dfrac{1}{8}$

2. $\dfrac{1}{3}$

3. $\dfrac{3}{20}$

4. $\dfrac{2}{7}$

5. $\dfrac{2}{5}$

EXERCISE 4-5

Answer the following questions about percents.

1. What is 30% of 520?

2. What is 28% of 10?

3. What is 10% of 28?

4. What is 6.5% of 2?

5. Andy invited 70% of the 20 students in his class to his birthday party. How many students did he invite? How many did he *not* invite?

EXERCISE 4-6

Let's put what we've learned about interest to use in answering the following questions.

1. How much interest is earned in a year on a deposit of $200 if the interest rate is 4%?

2. How much interest is earned in 3 months on a deposit of $400 if the interest rate is 3%?

3. How much interest does Kathy pay on a $1,000 loan if the interest rate is 14% and she pays the loan back in one year?

4. Britt borrows $600 at an interest rate of 9%. He pays the loan back in two years. What is the total amount he paid?

EXERCISE 4-7

Apply what we've learned about percents to these everyday situations.

1. What is the new price of a $60 item that is discounted 25%?

2. What is the new price of a $42 item that is discounted 30%?

3. How much income tax does Amee pay if she earns $90,000 and her tax rate is 12%

4. How much sales tax will be added to an item marked $12 if the sales tax rate is 8%?

5. What is the total amount paid with a 15% tip on a bill of $21?

6. What is the total amount paid with a 20% tip on a bill of $40?

EXERCISE 4-8

Answer the following questions about percent increase and decrease.

1. Zane bought a house for $100,000 and sold it a few years later for $130,000. By what percent did the price of his house increase?

2. Nicholle got a raise at her job from $12 per hour to $14 per hour. By what percent did her salary increase?

3. Marisa walked six miles on Saturday, and John walked five miles. What percent greater was Marisa's walk?

4. There were approximately 50,000 grizzly bears in the United States in the early 1800s. Today, there are approximately 1,400. By what percent has the population of grizzly bears declined?

Ratios and Proportions

MUST ⚡ KNOW

⚡ A ratio compares two amounts and shows their relative sizes. A ratio can be increased or decreased by using a multiplier.

⚡ A proportion compares two equal ratios. We use proportions to help us increase or decrease amounts evenly.

⚡ Missing amounts in proportion or rate problems can be found by cross-multiplying.

A **ratio** compares two values and tells us how many there are of one thing compared to how many there are of another thing. In the last chapter we discussed percents, which are a type of ratio. A percent shows us how many of something per 100. A ratio is different from a fraction, which shows us how many parts there are out of the total number of parts. Ratios show us parts to parts rather than parts to whole.

A **proportion** compares two equal ratios. Proportions help us to increase or decrease a ratio. In this chapter, we will learn how to find amounts using ratios and proportions. We will also work with rates, which are a type of ratio.

Ratios

There are several different ways to express a ratio. The most common way to write a ratio is with a colon between the two values, like this 2:3. For example, if there is a ratio of two women to three men in a group, this can be written as:

- 2:3

- 2 to 3

- ratio of $\frac{2}{3}$ (also read as *ratio of two to three*)

Notice that the last option looks a lot like a fraction. A ratio is *not* a fraction, though, so if we express a ratio this way, we need to say that it is a ratio. Also notice that none of those options tells you the actual number of women or men in the group. None of them tells you the total number of people in the group, though we can say that there are five total parts to the ratio (two parts women and three parts men).

A ratio can tell us several things. In the example we just used, there are two parts women and three parts men. We can say:

- The ratio of women to men is 2:3.

- The ratio of men to women is 3:2.

- The ratio of women to the group total is 2:5.

- The ratio of men to the group total is 3:5.

You can see that the order you list the two values makes a big difference, so read questions carefully!

EXAMPLE

▶ Krishna is making a cake, and the recipe calls for three cups of flour and one cup of sugar. What is the ratio of flour to sugar?

▶ The question asks for flour to sugar, so be sure to put the amount of flour first. Three cups of flour to one cup of sugar can be written as a ratio of 3:1. Knowing this can help Krishna increase or decrease the recipe.

Ratios are also a way to express two values in their lowest, or simplest, form. If there are 20 zebras and 30 rhinos, we would say that the ratio of zebras to rhinos is 2:3. You can write the totals as a fraction and then reduce the fraction to find the ratio: $\dfrac{20}{30} = \dfrac{2}{3} = 2{:}3$.

EXAMPLE

▶ In a group of ten people, four are children and six are adults. What is the ratio of adults to children?

▶ This is a good example of the trickiness of writing a ratio. The numbers were listed with the children before the adults, but the question asks for the ratio of *adults to children*, so we need to put the number of adults first. Write the number of adults as the numerator of a fraction with the number of children as the denominator.

$$\frac{6\ adults}{4\ children}$$

▶ Now reduce the fractional ratio to its lowest form.

$$\frac{6}{4} = \frac{3}{2}$$

▶ The ratio of adults to children is 3:2.

In this example, the ratio of adults to children was $\frac{3}{2}$. We can also express that as a mixed number, $\frac{3}{2} = 1\frac{1}{2}$, so we can say that there are $1\frac{1}{2}$ (or 1.5) times as many adults as children.

Ratios are mostly used to determine the particular amount of something that is needed. For example, to make lemonade, we need 1 part lemon juice, 4 parts water, and $\frac{1}{2}$ part sugar. If we have 2 cups of lemon juice, the ratio will tell us how much water and sugar we need to add to make lemonade. If we know any one of the actual numbers, we can find them all. Here is what we know so far:

	Ratio	Actual
lemon juice	1	2 cups
water	4	
sugar	$\frac{1}{2}$	

We can multiply a ratio by any number to increase it. By what number do we need to multiply the lemon juice ratio in order to increase the ratio number (1) to the actual number (2)? We need to multiply 1 by 2 to get 2. The key is to multiply every part of the ratio by the same number to increase the ratio.

If we multiply the 1 cup for the lemon juice by 2, we also must multiply the 4 cups for the water by 2 and the 1/2 cup for the sugar by 2. If we don't increase every part of the ratio, our recipe will be off, and the lemonade will taste really weird. No one wants weird lemonade!

	Ratio	Multiplier	Actual
lemon juice	1	× 2	2 cups
water	4	× 2	8 cups
sugar	$\frac{1}{2}$	× 2	1 cup

Now we can make lemonade! What if we only had 1 cup of lemon juice? What would the multiplier be then? It's 1, so we can do this:

	Ratio	Multiplier	Actual
lemon juice	1	× 1	1 cup
water	4	× 1	4 cups
sugar	$\frac{1}{2}$	× 1	$\frac{1}{2}$ cup

What if we only had 1/2 cup of lemon juice? We can multiply the ratio by 1/2.

	Ratio	Multiplier	Actual
lemon juice	1	× $\frac{1}{2}$	$\frac{1}{2}$ cup
water	4	× $\frac{1}{2}$	2 cups
sugar	$\frac{1}{2}$	× $\frac{1}{2}$	$\frac{1}{4}$ cup

We can either increase a ratio or decrease it by using a multiplier.

BTW

Make sure you always use the same multiplier for the entire ratio. When a value is fixed and does not change, it is called a **constant**.

▶ Kalee is doing a chemistry experiment that calls for mixing 12 parts vinegar with 1 part baking soda. She has a 24-ounce bottle of vinegar. How many ounces of baking soda will she need if she uses the entire bottle of vinegar?

▶ Set up a chart like the one in the lemonade example. We only need two rows for the ingredients, though. Be careful to write the ratio in the same order it was given: 12 parts vinegar and 1 part baking soda.

	Ratio	Multiplier	Actual
vinegar	12		24 ounces
baking soda	1		

▶ What multiplier should we use to increase 12 to 24? A multiplier of 2. We should also increase the baking soda by a multiplier of 2. Since the actual amount of vinegar is in ounces, the baking soda will also need to be in ounces.

	Ratio	Multiplier	Actual
vinegar	12	× 2	24 ounces
baking soda	1	× 2	2 ounces

▶ Kalee, then, needs 2 ounces of baking soda.

You may also need to know about the total amounts, and we can use our chart for that too.

▶ There are four 7th graders for every five 8th graders at Lakeville Junior High. If the school has a total of 270 students, how many 8th graders are there?

▶ Saying *four 7th graders for every five 8th graders* is the same as saying the *ratio of 7th to 8th graders is 4:5*. Set up a chart and be careful to write the ratio in the same order it was given. Since the question involves the total, include a row for the totals.

	Ratio	Multiplier	Actual
7th graders	4		
8th graders	5		
Total			270

▶ If there are four parts 7th graders and five parts 8th graders, how many total parts are there to the ratio? $4 + 5 = 9$. Write that in the chart as the ratio total.

▶ Now we need to find the multiplier. The only actual number we have is the total number of students, so we will use that row to find the multiplier. Divide 270 by 9 to find the multiplier: $270 \div 9 = 30$.

	Ratio	Multiplier	Actual
7th graders	4		
8th graders	5		
Total	9	30	270

▶ Write the multiplier of 30 for that whole column in the chart. Since we only need the number of 8th graders, just multiply 5 by 30 to find the actual number of 8th graders.

	Ratio	Multiplier	Actual
7th graders	4	30	
8th graders	5	30	150
Total	9	30	270

▶ There are 150 8th graders. We could also multiply 30 times 4 to get the number of 7th graders (120), but that isn't what the question asked. So we don't really need to spend time finding that.

▶ Always be sure to double-check the wording of the question so that you answer what is needed and not what isn't.

Proportions

A proportion is a set of two equal ratios. Two proportional ratios increase or decrease at the same rate. The *ratio* row and the *actual* row of our chart are equal ratios and can be called proportional. The multiplier column is the same all the way across; it is a constant. For proportions, we call the multiplier the **constant of proportionality**. In the ratio charts we did, we used the ratio and actual amount to find the multiplier, which is the constant of proportionality, so you already know how to do that!

Like our ratio chart, proportions are also used to find missing amounts. If we know one of the ratios and one actual amount, we can find the other actual amount. A proportion is written as two equal fractions, like this:

$$\frac{1}{3} = \frac{3}{9}$$

In a proportion question, one of those numerators or denominators will be missing. We use a technique called **cross multiplying** to find the missing number. Have you noticed how, if we cross multiply these fractions, we get $1 \times 9 = 3 \times 3$? Both equal 9, right? If one of the numbers is missing, we can find it by writing an **equation**:

$$\frac{1}{3} = \frac{?}{9}$$

An equation is a mathematical statement that shows two things are equal. Since one of the numbers is missing, we need to call it something, so we use a **variable**, which is a letter that stands in for an unknown number. Instead of writing question marks, let's call the missing number y:

$$\frac{1}{3} = \frac{y}{9}$$

Doing cross multiplication gives us $1 \times 9 = 3 \times y$. We can simplify that to $9 = 3y$. Since we need to find the value of y, we need to get the y by itself on one side of the equals sign. We can do that here by dividing both sides of the equation by three.

$$9 = 3y$$
$$\frac{9}{3} = \frac{3y}{3}$$
$$3 = y$$

Let's see how to use proportions to find missing amounts. Once you get the hang of it, it will be even easier than using a ratio chart.

EXAMPLE

▶ Nick wants to give every guest at his party the same number of cookies. If there are six guests and 30 cookies, how many cookies can he give each guest?

▶ Set up the proportion using the numbers given. Let's use c for the number of cookies each guest will get. Be careful to put the amounts for the second ratio in the same order as the first.

$$\frac{6 \text{ guests}}{30 \text{ cookies}} = \frac{1 \text{ guest}}{c \text{ cookies}}$$

▶ To find the missing number, cross multiply: $6 \times c = 30 \times 1$. To get c by itself, divide both sides by 6.

$$\frac{6c}{6} = \frac{30}{6}$$
$$c = 5$$

▶ Each guest gets 5 cookies.

You can set up a proportion to help you find any ratio equal to a ratio you already have.

▶ A dozen eggs costs $1.12. What is the cost for two dozen eggs? What is the cost for three dozen and four dozen?

▶ Set up the first ratio.

$$\frac{1\,\text{dozen}}{\$1.12} = \frac{2\,\text{dozen}}{d\,\text{dollars}}$$

▶ To find the missing number, cross multiply: $1 \times d = 1.12 \times 2$. Solve for d.

$$d = 1.12 \times 2 = 2.24$$

▶ You may have noticed that we are simply multiplying the price for one dozen by 2 to find the price for two dozen. To find the price for three dozen, we can multiply the price for one dozen by 3.

$$d = 1.12 \times 3 = 3.36$$

▶ To find the price for four dozen, we can multiply the price for one dozen by 4.

$$d = 1.12 \times 4 = 4.48$$

▶ All of these ratios are equal. Any of them can be reduced to the unit rate, the price per dozen.

$$\frac{1\,\text{dozen}}{\$1.12} = \frac{2\,\text{dozen}}{\$2.24} = \frac{3\,\text{dozen}}{\$3.36} = \frac{4\,\text{dozen}}{\$4.48}$$

Proportions can be used to increase the size of something equally. In geometry, we call proportional figures **similar**.

EXAMPLE

▶ The two triangles shown are similar. What is the length of side s?

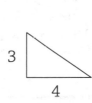 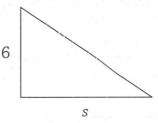

▶ Set up a proportion: $\dfrac{3}{4} = \dfrac{6}{s}$.

▶ Cross multiply: $3s = 24$.

▶ Divide both sides by 3 to solve for s: $\dfrac{3s}{3} = \dfrac{24}{3}$, so $s = 8$.

Look at the triangles in the example we just did. How much bigger is the second triangle? You can probably see that the second triangle is twice as big as the first. That means that the constant of proportionality is 2. Two times each of the dimensions of the first triangle gives us the dimensions of the second triangle.

Even without the pictures, we can determine the constant of proportionality as long as we know the triangles are similar triangles. Let's put the dimensions of the triangles into a chart.

Triangle 1	Constant of Proportionality	Triangle 2
left side: 3		left side: 6
bottom side: 4		bottom side: s

Just as we did to find the multiplier in the ratio charts, we can ask ourselves, *3 times what equals 6?* It's 2. That means the constant of proportionality is 2, and we can multiply the bottom side of the left-hand triangle, 4, by 2 to find the bottom side of the triangle on the right.

Triangle 1	Constant of Proportionality	Triangle 2
left side: 3	2	left side: 6
bottom side: 4	2	bottom side: 8

We can also use this knowledge to help us determine whether a set of values is proportional.

▶ Prices at a used book sale are shown in the chart here. Is the price proportional to the number of books purchased?

Books	Price
1	$2
2	$3
3	$5

▶ Determine the multiplier for one book: $1 \times \$2 = \$2$, so the multiplier is $2.

▶ Is the multiplier constant? For two books, it is: $2 \times \$2 = \$4$, but the chart says two books cost $3. That means the multiplier there is $1.50.

▶ The multiplier is not constant: $5 \times \$1 = \$5$, so the multiplier is different there, too.

▶ This means the relationship between the number of books and the price is *not* proportional. People are actually paying less per book when they buy more books!

IRL When we make a model of something, we are using proportions. If a room measures 18 feet by 12 feet, we can make a floor plan by drawing a proportionately much smaller room. $\frac{18}{12}$ is a fraction that can be reduced all the way down to $\frac{3}{2}$, so we can choose the scale we like.

We could make a model that is $\frac{18\,cm}{12\,cm}$ or $\frac{9\,cm}{6\,cm}$ or $\frac{6\,in.}{4\,in.}$ or $\frac{3\,in.}{2\,in.}$ We could even make a 3D model of our entire house this way! This is called making a scale model.

All we have to do is decide on a constant of proportionality and then multiply all the dimensions by that constant. We will practice doing this in Chapter 12.

Rates

A **rate** is a ratio that compares two amounts with different units, such as cups and gallons. A **unit rate** shows the ratio between an amount and *one unit* of the other, such as miles per hour. Rates can be scaled up or down, just like proportions. If a person drives 240 miles in three hours, the rate can be expressed as:

$$\frac{240\ \text{miles}}{3\ \text{hours}} \quad \text{or} \quad \frac{120\ \text{miles}}{1.5\ \text{hours}} \quad \text{or} \quad \frac{80\ \text{miles}}{1\ \text{hour}} \quad \text{or even} \quad \frac{1.3\ \text{miles}}{1\ \text{minute}}$$

Proportions can be used to solve rate problems.

EXAMPLE

▶ Lori runs six miles in 72 minutes. How far could she run, at the same pace, in 120 minutes?

▶ Set up a proportion. Let's use *d* for the unknown distance. Keep the units consistent.

$$\frac{6\ \text{miles}}{72\ \text{minutes}} = \frac{d}{120\ \text{minutes}}$$

▶ Cross multiply: $6 \times 120 = 72d$, so $720 = 72d$.

▶ Divide both sides of the equation by 72 to find *d*.

$$\frac{720}{72} = \frac{72d}{72}$$

▶ $d = 10$. Lori can run ten miles in 120 minutes.

Proportions can also be used to find the unit rate. The unit rate is how much of something per one unit.

▶ Lori runs six miles in 72 minutes. What is her rate in miles per hour?

▶ Set up a proportion. Be careful about units. The question gives Lori's distance in minutes but asks about hours. Let's use 60 minutes instead of 1 hour to have consistent units. Again, we can use d for the unknown distance.

$$\frac{6\,\text{miles}}{72\,\text{minutes}} = \frac{d}{60\,\text{minutes}(1\,\text{hour})}$$

▶ Cross multiply: $6 \times 60 = 72d$, so $360 = 72d$.

▶ Divide both sides of the equation by 72 to find d.

$$\frac{360}{72} = \frac{72d}{72}$$
$$d = 5$$

▶ Lori can run five miles in one hour (60 minutes), so her rate is 5 miles per hour. We can also write that as 5 mph.

IRL Unit rate is very helpful in the grocery store. How do you know which is more expensive: 12 ounces of olive oil for $7.99 or 30 ounces of olive oil for $17.99? You can't compare them directly since they aren't the same size.

Many stores now show the unit rate to help customers compare the prices of items of different amounts. You may see a listing on the shelf that shows the price per ounce: $0.67 per ounce for the 12-ounce bottle and $0.60 for the 30-ounce bottle. The unit rate lets you know that the 30-ounce bottle is a better deal.

EXERCISES

EXERCISE 5-1

Solve the following ratio problems.

1. If there are 400 students and 32 teachers at a school, what is the ratio of students to teachers?

2. If there are 100 hot dogs and 80 hamburgers at a picnic, what is the ratio of hamburgers to hot dogs?

3. In a group of kittens, there are two white kittens and three tan kittens. What is the ratio of tan kittens to all kittens?

4. Bethesda High School has a student to teacher ratio of 20:1. If there are 640 students, how many teachers should there be?

5. A tray of 100 cookies has a ratio of 12 chocolate chip cookies to 8 sugar cookies. How many sugar cookies are there? How many chocolate chip?

6. Sergio is making shortbread cookies. The recipe says to use three parts flour, two parts butter, and one part sugar. Sergio has plenty of flour and sugar but only one cup of butter. If he uses the entire cup of butter, how much sugar and flour will he need?

EXERCISE 5-2

Solve these proportions for x.

1. $\dfrac{5}{25} = \dfrac{2}{x}$

2. $\dfrac{x}{3} = \dfrac{9}{54}$

3. $\dfrac{20 \text{ teachers}}{100 \text{ students}} = \dfrac{x \text{ teachers}}{30 \text{ students}}$

4. $\dfrac{104 \text{ apples}}{x \text{ oranges}} = \dfrac{13 \text{ apples}}{8 \text{ oranges}}$

Solve the following proportion word problems.

5. Landon needs to make 36 cupcakes. The recipe calls for two eggs and makes 24 cupcakes. How many eggs will Landon need to make 36 cupcakes?

6. If Molly lays bricks for two hours and uses 68 bricks, how many bricks will she use in three hours at the same speed?

7. If six inches is 15.24 centimeters, how many centimeters is eight inches?

8. A photo measures four inches wide and six inches long. To increase the photo proportionally to make it eight inches long, how many inches will the width need to be?

EXERCISE 5-3

Solve the following rate problems.

1. At a contest, Bruce ate 11 hot dogs in five minutes. How long would it take him to eat 20 hot dogs at the same rate?

2. Cheri's sewing machine can sew 1,100 stitches per minute. How many stitches can it sew in seven minutes?

3. Vicki earns $13.40 per hour at her job. If she works six hours on Tuesday, how much does she earn?

4. If the WINK Bakery factory can make 78 loaves of bread in two hours, what is the rate per hour?

5. Store A has a bag of 12 apples for $4. Store B has a bag of 20 apples for $9. What is the price per apple at each store? Which store has the better deal?

6. Ryan ran a 5-kilometer race in 23 minutes. Chris ran a 3-kilometer race in 17 minutes. Which runner was faster? What was that person's rate in minutes per kilometer?

The Coordinate Plane

MUST KNOW

 The coordinate plane is formed by the intersection of a horizontal number line and a vertical one.

 An ordered pair shows us the location of a point on the coordinate plane. Plotting ordered pairs allows us to graph lines, figures, and functions.

 A linear equation forms a straight line on a graph, for example, $x = 2y$. If the line goes through the origin, the relationship between x and y is considered proportional.

 A unit rate is a kind of ratio that shows us how many units of one quantity correspond to one unit of another quantity. The unit rate represents the slope of a line on the coordinate plane.

Perpendicular lines intersect each other at a 90° angle. The **coordinate plane** is a two-dimensional plane formed by the intersection of two perpendicular lines: a horizontal line, which is called the *x*-**axis**, and a vertical line, which is called the *y*-**axis**. The point where the two lines intersect is called the origin. The coordinate plane looks like this:

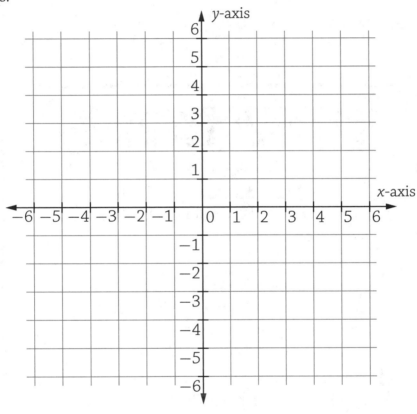

We call this a **graph**, and it is used for showing lines and geometric figures and functions. Each axis is divided into points, as on a number line. Look at the *x*-axis. It has positive numbers to the right of the origin and negative numbers to the left of the origin. We usually show the coordinate plane with an interval of one between each number marked, so the numbers increase by one as we move to the right along the *x*-axis. The *y*-axis is done the same way, but with the positive numbers increasing as we go up from the origin and negative numbers decreasing as we go down from the origin.

The coordinate plane is divided into quadrants, or fourths, as shown in the next graph. Each section is numbered using Roman numerals. Quadrant I (or the first quadrant) contains points with positive coordinates for both x and y. Quadrant II (or the second quadrant) contains points with negative x coordinates and positive y coordinates. Quadrant III (or the third quadrant) contains points with negative coordinates for both x and y. Quadrant IV (or the fourth quadrant) contains points with positive x coordinates and negative y coordinates. The divisions look like this:

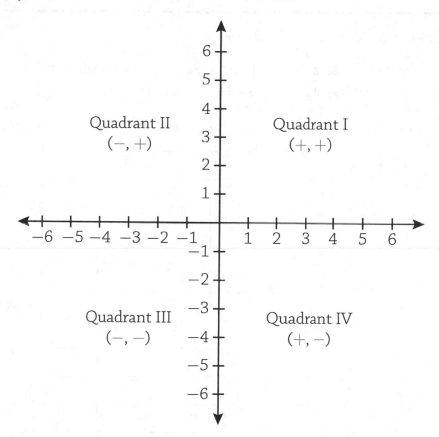

Plotting Ordered Pairs

An **ordered pair** is the set of x and y coordinates for a given point on the coordinate plane. Since the origin is the intersection of the lines, we call that point (0, 0). The first number listed in the pair is the x coordinate, and the second number is the *y* coordinate. We can find the point on a graph by going out from the origin along the *x*-axis to the location of the *x* coordinate and then going up or down (depending on whether the coordinate is positive or negative) to the location of the *y* coordinate. For the point (7, −1), $x = 7$ and $y = -1$. We can find the point on the graph by going right from the origin to 7 and then down along the *y*-axis to −1. If we draw a dot where the point is located, this is called **plotting** a point.

▶ Plot point (2, 3) on a coordinate plane.

▶ Since the coordinates for this point are (2, 3), we find positive 2 on the *x*-axis. Then we move up to positive 3 on the *y*-axis.

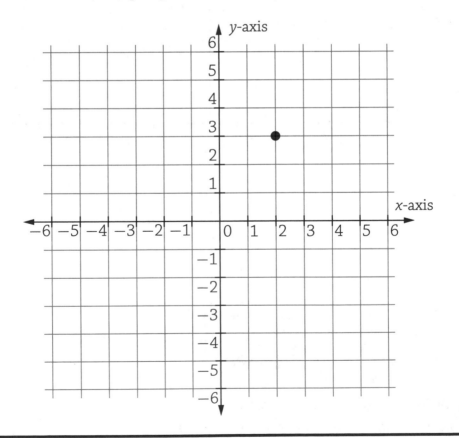

For a point with negative coordinates, we go to the *left* along the x-axis and *down* along the y-axis.

▶ Plot point (−4, −2) on a coordinate plane.

▶ Since the coordinates for the point are (−4, −2), we find −4 on the x-axis. Then we move down to −2 on the y-axis.

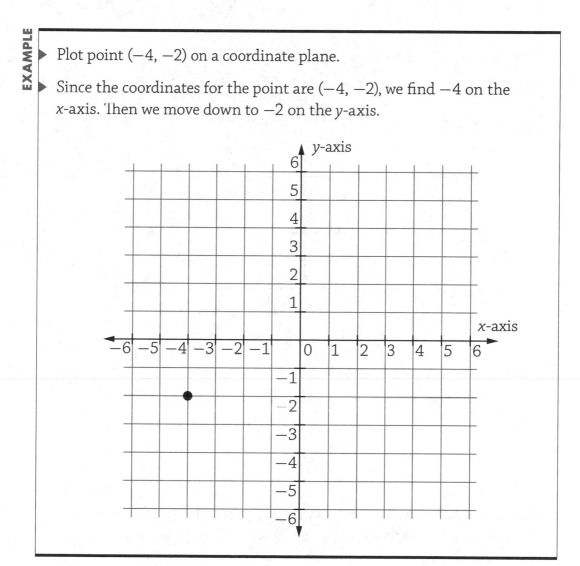

To find the coordinates of a point on a graph, we use the same ideas. Count how many spaces to the right or left along the x-axis the point is, and that is the *x* coordinate. Count how many spaces up or down along the *y*-axis the point is, and that is the *y* coordinate.

▶ What are the coordinates of points *A*, *B*, and *C*?

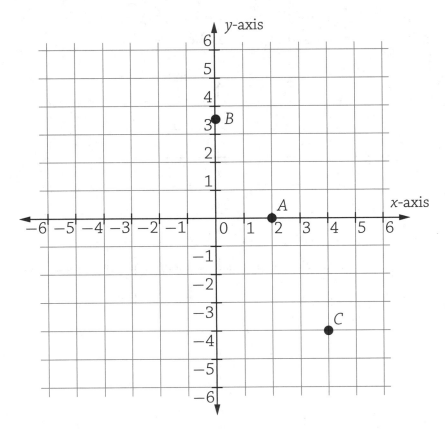

▶ Find point *A*. It is at 2 on the *x*-axis, so the *x* coordinate is 2. Where is it on the *y*-axis? It's on the line, so it is at 0 and the *y* coordinate is 0. Point *A* is at (2,0).

▶ Find point *B*. It is on the *x*-axis, so the *x* coordinate is 0. Point *B* is at 3.5 on the *y*-axis, so the *y* coordinate is 3.5. Point *B* is at (0, 3.5).

▶ Find point *C*. It is at 4 on the *x*-axis, so the *x* coordinate is 4. Point *C* is at −4 on the *y*-axis, so the *y* coordinate is −4. Point *B* is at (4, −4).

Graphing Relationships

We can use graphing to show us relationships between values for x and y. Using values of x and y, we can plot several of the possible points and then connect them with a line. This can help us visualize the relative values.

Let's find the graph for $x = y + 1$. The first thing we need to do is find some possible values for x and y. Let's make a chart. We want to use a range of values to see what the line does, so let's fill in some positive and negative values for y and then see what x is for each of them. This is called a **function table**.

x	y
	−2
	−1
	0
	1
	2

Now we can use the equation $x = y + 1$ to fill in the values for x. For example, if $y = 1$ and $x = y + 1$, then $x = 2$.

x	y
−1	−2
0	−1
1	0
2	1
3	2

For each pair of values, we can plot the point on a graph. We need to plot the points (−1, −2), (0, −1), (1, 0), (2, 1), and (3, 2).

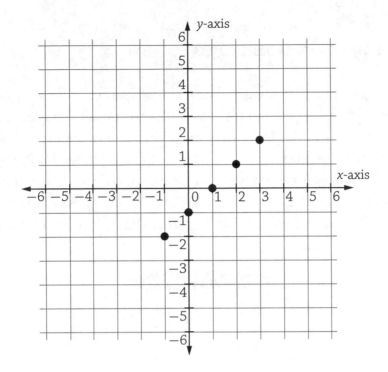

Now, we can draw a line between all the points that we plotted:

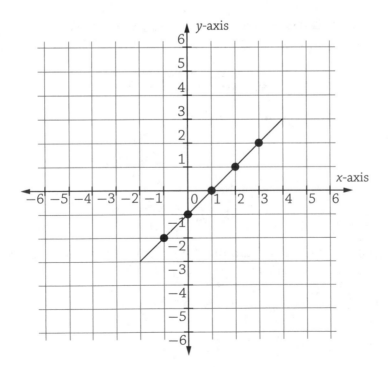

EXAMPLE

▶ Draw the graph for $x = 2y$.

▶ First, we find some points on the graph. Let's draw a function table; choose a few values for y and then see what x is for those values. Be sure to use both positive and negative values for y.

x	y
−2	−1
0	0
2	1

▶ Now graph those points and draw a line connecting them.

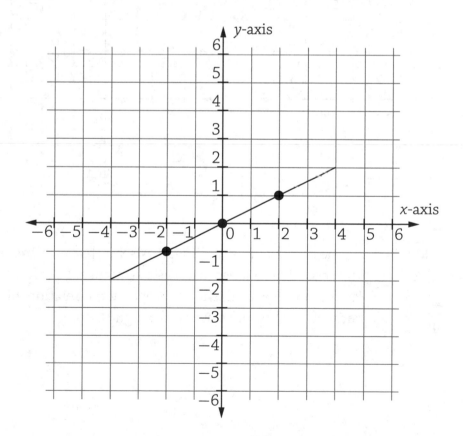

▶ Notice that for this graph, the line goes through the origin. That means that the relationship between x and y is **proportional**. In this case, $x = 2y$.

For the last example, we knew there was a proportional relationship between x and y before we drew the graph. We can go the other way as well and determine whether a graph is of a proportional relationship, and, if so, we can determine the constant of proportionality. A straight line is called a **linear equation**, and one that goes through the origin is proportional.

Look at these two graphs. Which shows a linear equation?

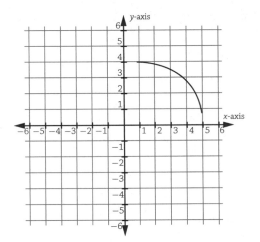
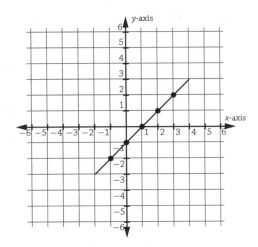

The first graph is not linear. We call that **nonlinear**. The second graph, because it is a straight line, is linear. It does not, however, show a proportional relationship because the line does not go through the origin.

▶ What is the constant of proportionality for the linear equation shown here? Write the equation of the line.

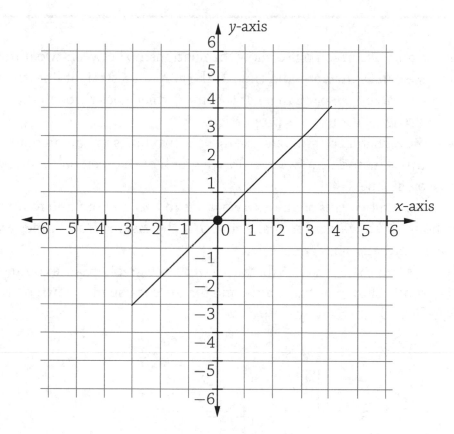

▶ From the graph, we can construct the function table.

x	y
−2	−2
−1	−1
0	0
1	1
2	2

▶ For every value, $x = y$. The constant of proportionality, therefore, is 1. The equation of the line is $x = y$.

We can also use graphing to show the relationship between two different quantities, such as distance and time. If the graph is a straight line, then the rate is constant (does not change). If the line formed goes through the origin (0, 0), then the relationship is proportional.

If we know that a certain person can walk two miles in half an hour and four miles in one hour, we can draw a graph of that person's walking distance over time.

We are given two sets of values: two miles in half an hour and four miles in one hour. First, we need to make sure that the distances and times are on the same scale, and they are.

Now let's draw the graph. A distance-over-time graph is quite common; time is usually shown on the x-axis, and distance is usually shown on the y-axis, so let's do it that way.

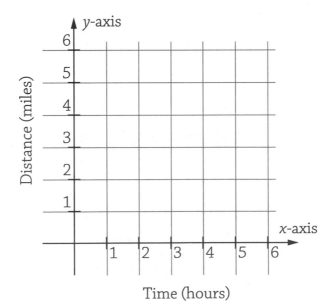

Now we can use the values we know as x and y coordinates for points on the graph: two miles in $\frac{1}{2}$ hour becomes the point $(\frac{1}{2}, 2)$ and four miles in one hour becomes the point $(1, 4)$. Let's plot those points and then draw a line connecting them.

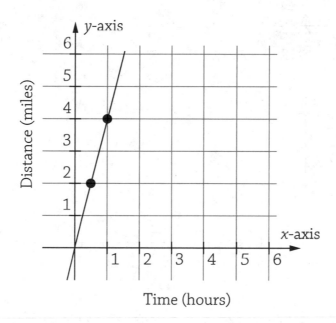

Time (hours)

Drawing the line on the graph allows us to do several things. Since the line is straight, we can conclude that the person's walking speed is constant. Since the line goes through the origin, we can conclude that the relationship between distance and time is proportional. We can also use the graph to predict other values. For example, in 1.5 hours, we can see that the person would walk six miles at the same rate.

▶ The next graph shows the distance traveled by a helicopter over time. What is the speed of the helicopter in miles per hour?

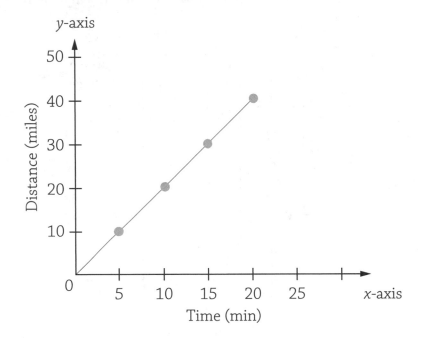

▶ First, note what the graph shows. This is a graph of distance in miles over time in minutes. Since we are asked to find the speed in miles per *hour*, we need to pay attention and convert the minutes to hours.

▶ Let's identify the points shown: (5, 10), (10, 20), (15, 30), and (20, 40). That means the helicopter goes 10 miles in five minutes, 20 miles in ten minutes, 30 miles in fifteen minutes, and 40 miles in twenty minutes.

▶ Since the line is straight and goes through (0, 0), we know that the speed is constant and that distance and time are proportional. We can use any of the points to set up a proportion and find the distance per one hour. Remember to convert the minutes to hours!

▶ Let's use the 40 miles in twenty minutes point for our proportion. We need to express twenty minutes in hours. An hour is sixty minutes, so twenty minutes is $\frac{1}{3}$ of an hour. Set up the proportion:

$$\frac{40\,\text{miles}}{\frac{1}{3}\,\text{hour}} = \frac{x\,\text{miles}}{1\,\text{hour}}$$

▶ Cross multiply to find x: $\frac{1}{3}x = 40 \times 1$. Multiply both sides of the equation by three to get x by itself: $x = 120$. The helicopter is flying at a constant speed of 120 miles per hour.

We can even compare two rates on the same graph. All we need to do is plot both sets of points on the same graph. Most comparison graphs will use a different color or a different type of line for each set of values. This requires that the graph have a **legend**, which shows what color or type of line corresponds to each set of values.

EXAMPLE

▶ Look at the graph below. How much farther is Valerie than Kai after four hours?

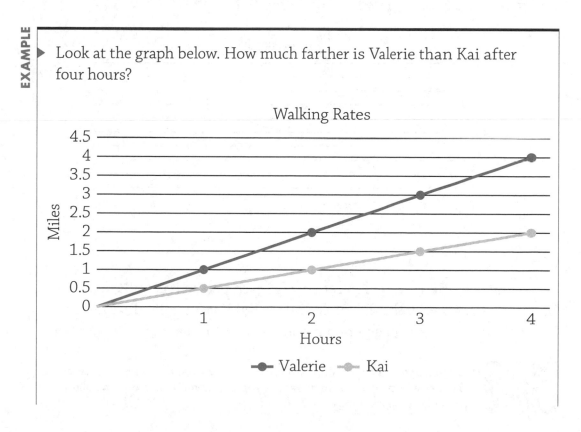

> We can easily see from the graph that Valerie is walking faster than Kai. The question asks us to compare their distance after 4 hours, so let's see where they are. Valerie has walked 4 miles after 4 hours. Kai has walked 2 miles after 4 hours. That means Valerie is 2 miles farther than Kai at that time.

Unit Rate

Graphing rates on the coordinate plane also allows us to easily see the **unit rate**. A unit rate is the amount per one unit of something, such as miles per hour or number of cookies per person. Let's look at the graph we just used and determine the unit rate for Valerie and Kai.

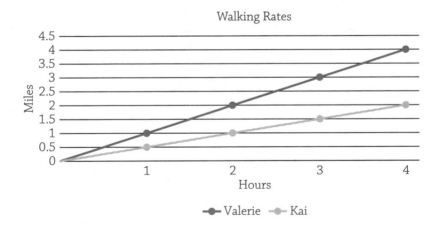

To find the unit rate, we look at how far each person has walked after one hour. For Valerie, that is 1 mile and for Kai it is 0.5 mile. That means Valerie's unit rate is 1 mile per hour and Kai's unit rate is 0.5 miles per hour.

 IRL We often use graphs like the previous one to show trends over time, such as the change in the price of something. Graphs with more than one line allow us to quickly compare things.

EXAMPLE

> Use the next graph to find the unit rate for each person in a hot dog eating contest.

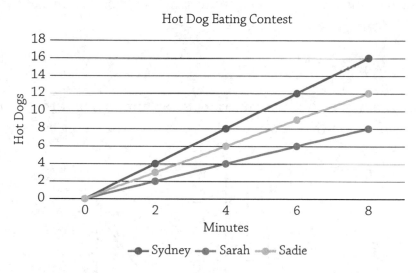

> To find the unit rate, we need to know the number of hot dogs each person eats in one minute. For Sydney, that is two hot dogs per minute, and for Sarah, that is one hot dog per minute.

> It's a little difficult to see exactly how many hot dogs Sadie eats per minute, but since the rates are constant, we can look at any number of minutes we like and then do a proportion to determine the unit rate. Look at Sadie's value for four minutes. She eats six hot dogs in four minutes.

$$\frac{6 \text{ hot dogs}}{4 \text{ minutes}} = \frac{x \text{ hot dogs}}{1 \text{ minute}}$$

> So $4x = 6$ and $x = \dfrac{6}{4} = \dfrac{3}{2} = 1\dfrac{1}{2}$, so that is $1\dfrac{1}{2}$ hot dogs per minute for Sadie.

The unit rate on a graph is the same as the **slope** of the line on the coordinate plane. Slope is a measure of how much the line rises or falls as

we look at the coordinate plane from left to right. We can find the slope
of a line on a coordinate plane by counting how far the line goes vertically
between two points on the line—the **rise**—divided by how far the line goes
horizontally—the **run**—between those same two points. We call this *rise
over run*. We can therefore write:

$$\text{slope} = \frac{\text{rise}}{\text{run}}$$

Picture a series of similar right triangles forming a line with positive
slope. Let's measure the base and height of one of the triangles here:

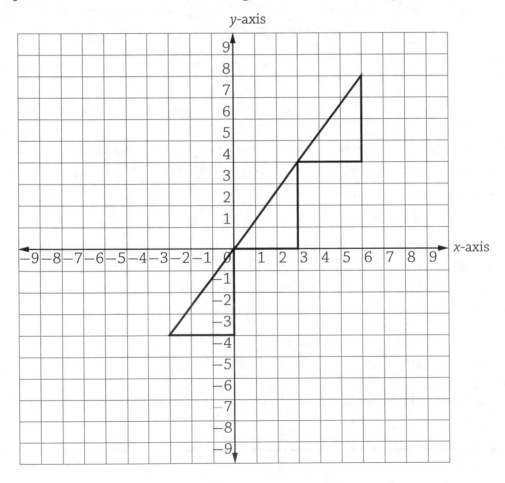

The rise, or the height of the triangle we made is 4 and the run, or the
length of the base of our triangle, is 3.

For this line, the slope is: $\dfrac{\text{rise}}{\text{run}} = \dfrac{4}{3}$. The unit rate for this graph is the same as the slope: $\dfrac{4}{3}$.

If the line goes up as we look at the graph from left to right, the slope is positive. If the line goes down as we look at the graph from left to right, the slope is negative.

EXAMPLE

▶ Find the unit rate of the line on the below graph by determining its slope.

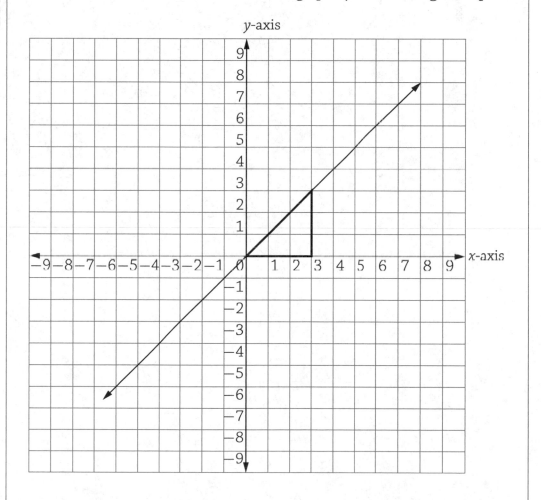

▶ Draw a right triangle forming part of the line. (I've already done it for you here). Measure the base and height of the triangle. The base is 3 and the height is also 3. The slope is the rise over the run.

$$\text{slope} = \frac{\text{rise}}{\text{run}} = \frac{3}{3} = 1$$

▶ That means that the unit rate is 1.

EXERCISES

EXERCISE 6-1

Draw a graph with x- and y-axes, and plot the following points on the graph.

1. (3, 2)

2. (−1, 4)

3. (2, −3)

4. (−2, −4)

5. (0, 3)

6. (3, 0)

Refer to the graph below to answer questions 7 and 8.

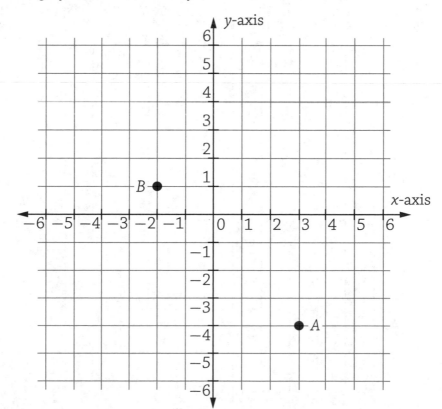

7. What are the coordinates of point A shown in the graph? In which quadrant is point A?

8. What is the x-coordinate of point B shown in the graph? In which quadrant is point B?

EXERCISE 6-2

Follow the instructions for each of the following questions.

1. Fill out a function table with x and y columns and draw the graph for $x = y - 1$. Is it linear? Are x and y proportional?

2. Fill out a function table with x and y columns and draw the graph for $x = 3y$. Is it linear? Are x and y proportional?

3. What is the equation of the line shown here? What is the constant of proportionality? You can make a function table to help you.

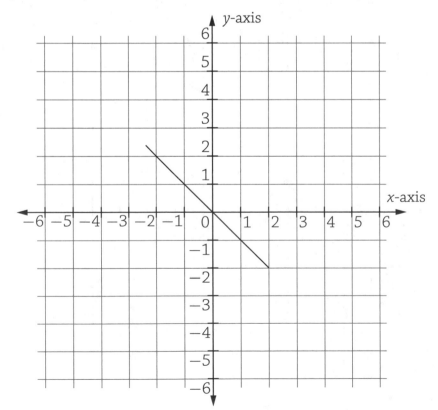

EXERCISE 6-3

Use the graph below for questions 1–2.

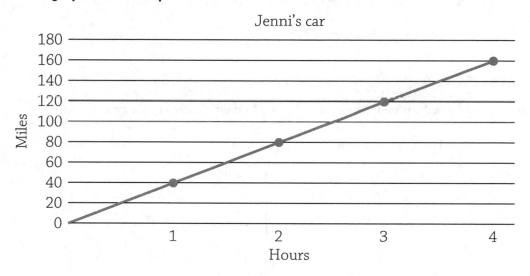

1. How far has Jenni's car gone after $2\frac{1}{2}$ hours?

2. How long does it take the car to go 40 miles?

Use this graph for questions 3–5:

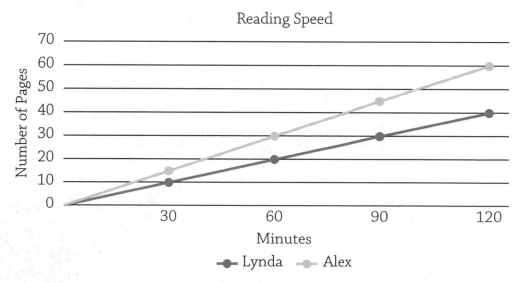

3. How many pages has Lynda read in one hour?

4. How long does it take Alex to read 45 pages?

5. Who is reading at a faster rate?

EXERCISE 6-4

Let's apply what we've learned about rates to the next two questions.

1. What is the unit rate for the graph here?

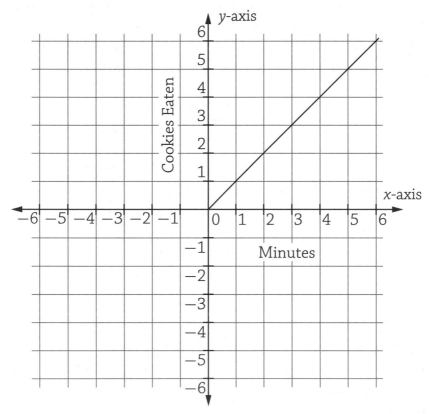

2. Draw a graph that shows a person walking at 0.50 mile per hour.

 **Exponents and Roots**

 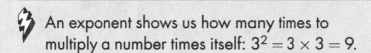

s we move more into algebra, we will have to learn how to deal with more complicated numbers. These include working with exponents and roots. You have probably seen a number with an exponent, such as 2^2, and a number with a root, for example, $\sqrt{25}$. In this chapter, we will learn how to perform operations on these types of numbers. We will also learn to use scientific notation, which uses exponents on powers of 10 to express very large or very small numbers.

Exponents

An **exponent** tells us how many times to multiply a number by itself. The number we are multiplying is called the **base**, and the exponent is written as a smaller number to the upper right of the base, like this:

The number 3 is the base and the little number 2 is the exponent. The exponent tells us that we need to multiply the number 3 two times, so 3×3. An exponent is sometimes called a **power**, and we might, for 3^2, say *raise 3 to the second power*.

For 3^4, where the exponent is a 4, we would multiply the number 3 four times, so: $3 \times 3 \times 3 \times 3$. We can also *say raise 3 to the fourth power*. Using an exponent is really just a short way to write these longer strings of multiplication. Instead of writing "$3 \times 3 \times 2 \times 2 \times 2 \times 2$," we can just write: $3^2 \times 2^4$.

Remember how zero is special? That applies to exponents as well. Any number raised to the 0 power is 1. That means 5^0 equals 1 and $1{,}283^0$ equals 1. We will see why when we discuss division.

The number 1 is also a little special. One raised to any power is one, because 1 times 1 equals 1 no matter how many times you do it: 1^3 equals 1 and 1^{143} equals 1. Any number raised to the first power is just the number itself: 5^1 equals 5 and $1{,}283^1$ equals $1{,}283$.

Adding and Subtracting with Exponents

Since numbers with exponents represent multiplication, they cannot be directly added or subtracted: $3^2 + 3^2 \neq 6^2$. You can always write out the exponents to see what happens. Let's see:

$$(3 \times 3) + (3 \times 3) = 9 + 9 = 18$$
$$6^2 = 6 \times 6 = 36$$
$$18 \neq 36$$

The same thing happens with subtraction:

$$3^2 - 2^2 \neq 1^2$$
$$(3 \times 3) - (2 \times 2) = 9 - 4 = 5$$
$$1^2 = 1 \times 1 = 1$$
$$5 \neq 1$$

The only way to add or subtract exponents is to calculate the values first:

$$3^2 - 2^2 = 9 - 4 = 5$$

Multiplying with Exponents

The good news is that, since numbers with exponents represent multiplication, they *can* be multiplied and divided, as long as the base is the same.

What is 3^2 times 3^2? Let's expand it and see: $(3 \times 3) \times (3 \times 3) = 9 \times 9 = 81$. There is a shortcut we can use here: $(3 \times 3) \times (3 \times 3)$. This is the same as: $3 \times 3 \times 3 \times 3$, which is the same as 3^4. That means: $3^2 \times 3^2 = 3^4$. When you multiply numbers with the same base, you can just *add* the exponents: $3^2 \times 3^2 = 3^{2+2} = 3^4$.

This does not work with different base numbers: 3^2 times 2^2 does not equal 6^4 because: $(3 \times 3) \times (2 \times 2) = 9 \times 4 = 36$ and $6^4 = 6 \times 6 \times 6 \times 6 = 1,296$. You can still calculate 3^2 times 2^2, but you can't use the shortcut of adding the exponents. Do that only when the base numbers are the same.

▶ What is 5^3 times 5^2?

▶ Since we are multiplying numbers with the same base, we can use the shortcut to add the exponents.

$$5^3 \times 5^2 = 5^{(3+2)} = 5^5$$

▶ If you ever want to check your work, just expand it.

$$5^3 \times 5^2 = (5 \times 5 \times 5) \times (5 \times 5) = 5 \times 5 \times 5 \times 5 \times 5 = 5^5$$

Dividing with Exponents

Let's see what happens when we divide numbers that have exponents.

What is 8^4 divided by 8^2? Let's expand it and see: $\dfrac{8 \times 8 \times 8 \times 8}{8 \times 8}$. We can reduce that before we try to calculate:

$$\frac{8 \times 8 \times \cancel{8} \times \cancel{8}}{\cancel{8} \times \cancel{8}} = \frac{8 \times 8}{1} = 64$$

We can use a shortcut here: $\dfrac{8^4}{8^2} = 8^{(4-2)} = 8^2$. When you divide numbers with the same base, you can just *subtract* the exponents!

Again, this does *not* work with different base numbers: 4^2 divided by 2^2 does not equal 2 because: $(4 \times 4) \div (2 \times 2) = 16 \div 4 = 4$. You can still calculate 4^2 divided by 2^2, but you can't use the shortcut of subtracting the exponents. Do that *only* when the base numbers are the same.

▶ What is 11^5 divided by 11^4?

▶ Since we are dividing numbers with the same base, we can use the shortcut to subtract the exponents → $11^5 \div 11^4 = 11^{(5-4)} = 11^1 = 11$.

▶ If you ever want to check your work, just expand it out.

$$\frac{11 \times \cancel{11} \times \cancel{11} \times \cancel{11} \times \cancel{11}}{\cancel{11} \times \cancel{11} \times \cancel{11} \times \cancel{11}} = \frac{11}{1} = 11$$

Remember the zero exponent rule? Any number to the zero power is equal to one. Division can show us why that is true. When we divide numbers with exponents, we subtract the exponents. What happens if we divide 4^5 by 4^5?

$$\frac{4^5}{4^5} = 4^{5-5} = 4^0$$

Any fraction with equal numerator and denominator is equal to one, right? For example, $\frac{2}{2}$ equals 1. This is true for $\frac{4^5}{4^5}$ as well. $\frac{4^5}{4^5}$ equals 1 and, as we just saw, $\frac{4^5}{4^5}$ equals 4^0, so 4^0 equals 1.

BTW

Remember, you cannot do any operations on exponents with different base numbers. You cannot add, subtract, multiply, or divide 4^2 and 3^7 unless you calculate the values without the exponents: $4^2 = 16$ and $3^7 = 2,187$.

Raising an Exponent to a Power

We can even take a number with an exponent and raise the whole thing to a power, like giving the exponent an exponent. That looks like this: $(3^2)^2$.

To calculate that, we can expand it to see what happens: $(3^2)^2 = (3^2) \times (3^2) = 3 \times 3 \times 3 \times 3 = 3^4$. When we raise a power to a power, it is like multiplying the exponents: $(3^2)^2 = 3^{2 \times 2} = 3^4$.

▶ What is $(8^4)^3$?

▶ Since we are raising a number with an exponent to a power, we can use the shortcut to multiply the exponents → $(8^4)^3 = 8^{(4 \times 3)} = 8^{12}$.

▶ If you ever want to check your work, just expand it out.

$$(8^4)^3 = (8^4) \times (8^4) \times (8^4) = (8 \times 8 \times 8 \times 8)$$
$$\times (8 \times 8 \times 8 \times 8) \times (8 \times 8 \times 8 \times 8) = 8^{12}$$

What should you do if you see a number like 8^{4^3}? Is that the same thing as $(8^4)^3$? No. We have to follow the order of operations (PEMDAS). There are no parentheses, so we need to do the exponent first. Calculate: $4^3 = 64$. What we really have then is 8^{4^3} equals 8^{64}. This is not common, however, so don't lose sleep over it!

What if we have more than one item raised to a power? How about $(8x)^2$? Let's expand it and see what happens: $(8x)^2 = (8x) \times (8x)$. Since we can multiply in any order, we could write this as $(8)(x)(8)(x)$, or we can write it as $(8)(8)(x)(x)$. That's the same as $(8^2)(x^2)$. In other words, we distribute the exponent across all the items in parentheses: $(8x)^2 = 8^2 \times x^2$.

▶ What is $(4x^3 3y^2)^3$?

▶ We need to distribute the exponent outside the parentheses (3) across all the items inside the parentheses.

$$(4x^3 3y^2)^3 = (4^3) \times (x^3)^3 \times (3^3) \times (y^2)^3$$

▶ Now multiply the exponents for $(x^3)^3$ and $(y^2)^3$.

$$(4x^3 3y^2)^3 = (4^3) \times (x^9) \times (3^3) \times (y^6)$$

▶ Now we can write it without the parentheses to simplify.

$$(4x^3 3y^2)^3 = 4^3 x^9 3^3 y^6$$

Negative Numbers

Everything we have done so far has been with positive numbers. What about negative numbers? The same principles apply if the base number is negative: $(-4)^3 = (-4)(-4)(-4) = -64$.

- To multiply, you add the exponents: $(-2^4)(-2^3) = (-2)^7$.

- To divide, you subtract the exponents: $\dfrac{-2^6}{-2^3} = (-2)^3$.

- To raise a negative base with an exponent to a power, multiply the exponents: $(-2^3)^2 = -2^6$.

What if the exponent itself is negative? What is 5^{-2}? Negative exponents are the multiplicative inverse of the positive exponent, so: $5^{-2} = \dfrac{1}{5^2} = \dfrac{1}{25}$.

Anytime you see a negative exponent, just rewrite it as a fraction with 1 on top and the positive version of the exponent on the bottom.

EXAMPLE

▶ What is $(6^{-5})^3$ as a fraction?

▶ First, rewrite the negative exponent:

$$(6^{-5})^3 = \left(\frac{1}{6^5}\right)^3$$

▶ Now distribute the power of 3 to the entire fraction. Remember that when you raise an exponent to a power, you multiply the exponents together:

$$\left(\frac{1}{6^5}\right)^3 = \frac{1^3}{6^{15}} = \frac{1}{6^{15}}$$

▶ That's some high school–level math you're doing here. You rock!

IRL Exponents are used often in many fields such as architecture, finance, biology, and construction—just to name a few. For example, to calculate compound interest on a loan, we use the formula $P\left(1 + \frac{r}{n}\right)^{nt}$ where P is the principle amount, r is the interest rate, n is the number of times interest is compounded per unit t, and t is the time.

Roots

As we just saw in our discussion of exponents, x^2 means x times x. Raising a number to the second power is also called **squaring** the number. The result of squaring is called a **perfect square**: $2^2 = 2 \times 2 = 4$. Four is a perfect square. Let's list a few perfect squares:

$$1^2 = 1 \qquad\qquad 6^2 = 36$$

$$2^2 = 4 \qquad\qquad 7^2 = 49$$

$$3^2 = 9 \qquad\qquad 8^2 = 64$$

$$4^2 = 16 \qquad\qquad 9^2 = 81$$

$$5^2 = 25 \qquad\qquad 10^2 = 100$$

The perfect square is the result of a number being multiplied times itself. The **square root** of a number is the number that was multiplied by itself to give us the square. In other words, for 4^2 equals 16, 4 is the square root of 16. We write a square root like this: $\sqrt{16} = 4$. The $\sqrt{}$ symbol is called a **radicand**. We know that 4 is the square root of 16 because $\sqrt{16}$ equals $\sqrt{4 \times 4}$. If you can break down the number under the radicand into two of the same number multiplied together, then you have a perfect square, and the number you have two of is the square root.

▶ What is $\sqrt{121}$?

▶ Let's see if we can break down the number 121. What times what gives us 121? It's bigger than 10 because 10 times 10 is 100. It's bigger than 10 times 11 because that would be 110. How about 11 times 11? Yes: $11 \times 11 = 121$!

▶ That means we can write $\sqrt{121} = \sqrt{11 \times 11}$, and $\sqrt{121}$ is a perfect square: $\sqrt{121} = 11$.

Unlike exponents, square roots cannot be negative. As you have learned, we can square a negative number. We can multiply: (-4) times (-4) equals 16 and say that $(-4)^2$ equals 16. Any number we square, positive or negative, results in a positive number. That means that when we take the square root of that square, we are taking the square root of a positive number. Negative numbers, then, do not have square roots: $\sqrt{16}$ equals only positive 4.

If you are presented with a larger number that you do not recognize as a perfect square, see if you can factor the larger number to find any perfect squares.

▶ What is $\sqrt{400}$?

▶ Let's see if we can break down the number 400. We can write it as 4 times 100, so: $\sqrt{400} = \sqrt{4 \times 100}$. Both 4 and 100 are perfect squares, so we can pull them both out $\rightarrow$ $\sqrt{4}$ equals 2 and $\sqrt{100}$ equals 10, so: $\sqrt{400} = \sqrt{4 \times 100} = 2 \times 10 = 20$.

What if the number under the radicand is *not* a perfect square? In that case, all we can do is try to reduce the number under the radicand as much as possible. We do this by pulling out any square roots contained in it: $\sqrt{32}$ equals $\sqrt{16 \times 2}$ and, since 16 is a perfect square, we can pull that root out and write it to the left of the radicand: $\sqrt{16 \times 2} = 4\sqrt{2}$.

When you have a number in front of the radicand, we call that a **coefficient**. What do we do with the $\sqrt{2}$? Nothing. It isn't a perfect square and it cannot be factored to give us a perfect square. Leave it as it is. If you have a calculator—many calculators have a root button—you can enter $\sqrt{2}$ to find out that it equals about 1.4, but that isn't necessary for the kinds of problems that you'll be doing at this point.

EXAMPLE

▶ What is $\sqrt{75}$?

▶ Let's see if we can break down the number 75. We can write it as 3 times 25, so: $\sqrt{75} = \sqrt{3 \times 25}$. We have one perfect square: 25.

▶ 3 is factored as small as possible and is not a perfect square, so we will just leave it under the radicand. Pull out the square root of the 25: $\sqrt{25}$ equals 5, so $\sqrt{75}$ equals $5\sqrt{3}$.

Estimating Roots

Without a calculator, it would be very difficult to find the value of $\sqrt{2}$ or any other root that is not the root of a perfect square. If you put $\sqrt{2}$ into your calculator, you will find that it is a decimal number that goes on forever without repeating. We call these numbers **irrational numbers**. You probably remember that a rational number is any number that can be written as a fraction with integers in the numerator and denominator.

Irrational numbers can't be written as fractions. When we need a value for an irrational root, we estimate. We can say that $\sqrt{2}$ equals about 1.4. We can write that with a sort of wavy equals sign that means *almost equal*: $\sqrt{2} \approx 1.4$. For many roots, we can estimate based on the closest perfect square: $\sqrt{3}$ is close to $\sqrt{4}$, which we know equals 2, so we can say $\sqrt{3}$ is a little less than 2. If we put $\sqrt{3}$ into a calculator, we can round the decimal number we get to 1.7.

EXAMPLE

▶ Estimate $\sqrt{30}$.

▶ We know that $\sqrt{25}$ equals 5 and $\sqrt{36}$ equals 6, so $\sqrt{30}$ will be somewhere in between 5 and 6.

▶ Since 30 is about halfway between 25 and 36, $\sqrt{30}$ should be about halfway between $\sqrt{25}$ and $\sqrt{36}$, so we can say: $\sqrt{30} \approx 5.5$.

▶ If we put $\sqrt{30}$ into a calculator, we get 5.48, so our estimate is pretty close!

Adding and Subtracting Roots

We cannot add or subtract numbers with exponents, but what about roots? They are a little different. We can directly add or subtract numbers that have the same root (the number under the radicand). We cannot add or subtract different roots without calculating them first:

$$3\sqrt{7} + 12\sqrt{7} = 15\sqrt{7}$$
$$7\sqrt{3} - 2\sqrt{3} = 5\sqrt{3}$$

We cannot directly add or subtract numbers like $2\sqrt{7}$ and $6\sqrt{5}$ because the roots are different.

EXAMPLE

What is $6\sqrt{7}$ minus $\sqrt{7}$?

Both terms are $\sqrt{7}$, so we can subtract directly. $\sqrt{7}$ is the same as $1\sqrt{7}$, so we are really subtracting this way: $6\sqrt{7} - 1\sqrt{7} = 5\sqrt{7}$.

If you are asked to add or subtract numbers with different roots, first check to see whether you can simplify either of the roots. You may find that they are the same after you simplify.

EXAMPLE

What is $5\sqrt{32} + 2\sqrt{50}$?

The roots look different, but are they? Let's see if we can reduce them: $5\sqrt{32} = 5\sqrt{16 \times 2}$. Since 16 is a perfect square, we can pull out the square root of 16, which is 4. Multiply the 4 times the coefficient 5 that is already outside the radicand to get $20\sqrt{2}$.

Now let's do $2\sqrt{50}$: $2\sqrt{50} = 2\sqrt{25 \times 2}$. Since 25 is a perfect square, we can pull out the square root of 25, which is 5. Multiply the 5 times the coefficient 2 that is already outside the radicand to get $10\sqrt{2}$.

Now we have: $20\sqrt{2} + 10\sqrt{2} = 30\sqrt{2}$. Magic!

Multiplying Roots

Just as with exponents, you can multiply roots easily. The only thing you have to keep in mind is that you need to treat coefficients and roots separately:

$$2\sqrt{5} \times 7\sqrt{3}$$

Multiply the coefficients: $2 \times 7 = 14$

Now multiply the roots: $\sqrt{5} \times \sqrt{3} = \sqrt{15}$

Put it all together: $14\sqrt{15}$

EXAMPLE

▶ What is $9\sqrt{3}$ times $2\sqrt{7}$?

▶ Since this is a multiplication problem, we don't need to worry about the numbers under the radicands being different. We just multiply them together (still under a radicand), and we multiply the two coefficients together.

$$9 \times 2 = 18$$

$$\sqrt{3} \times \sqrt{7} = \sqrt{21}$$

▶ Put 18 and $\sqrt{21}$ together, and we have $18\sqrt{21}$.

When you multiply the roots together, if you get a perfect square, you can pull out the root of that perfect square and make it a coefficient:

$$\sqrt{3} \times \sqrt{3} = \sqrt{9} = 3$$

If you already have a coefficient or two, you would just multiply the new one that you got from multiplying the roots to the others and multiply them all. Let's see how that works. Let's try $8\sqrt{12}$ times $\sqrt{3}$. Multiply the coefficients: $8 \times 1 = 8$. Now multiply the roots: $\sqrt{12} \times \sqrt{3} = \sqrt{36}$.

Since $\sqrt{36}$ is a perfect square, we can pull out the root 6, and it becomes a coefficient. Since we already have a coefficient of 8, we multiply: $8 \times 6 = 48$.

You may be wondering if we should have simplified $8\sqrt{12}$ before we did anything, but we don't have to. We can, but we don't have to. Use whichever method you find easier. Let's see the difference: $8\sqrt{12} \times \sqrt{3}$.

If we partly factor the 12 in $8\sqrt{12}$, we get $8\sqrt{4 \times 3}$. Since $\sqrt{4}$ is a perfect square, we can pull out the root 2, and it becomes a coefficient. Since we already have a coefficient of 8, we multiply: $8 \times 2 = 16$. Now we have: $16\sqrt{3} \times \sqrt{3} = 16 \times 3 = 48$. Same result!

▶ What is $5\sqrt{2}$ times $6\sqrt{80}$?

▶ Since this is a multiplication problem, we don't need to worry about the numbers under the radicands being different. We just multiply them (still under a radicand), and we multiply the two coefficients.

$$5 \times 6 = 30$$

$$\sqrt{2} \times \sqrt{80} = \sqrt{160}$$

▶ Can we simplify $\sqrt{160}$? Yes: $\sqrt{160} = \sqrt{16 \times 10}$. Since 16 is a perfect square, we can pull out the root of 16, which is 4. The remaining 10 cannot be broken down in a way that gives us a perfect square, so we just leave it alone, still under the radicand. Now we have: $30 \times 4\sqrt{10} = 120\sqrt{10}$.

Dividing Roots

Dividing roots is similar. Think of the two roots like a fraction that you may be able to simplify:

$$\sqrt{15} \div \sqrt{3} = \sqrt{\frac{15}{3}} = \sqrt{5}$$

If there are any coefficients, we have to divide those separately, just as we multiplied them separately when we multiplied numbers with roots:

$$4\sqrt{6} \div 2\sqrt{2} = \frac{4}{2}\sqrt{\frac{6}{2}} = 2\sqrt{3}$$

EXAMPLE

▶ What is $6\sqrt{15}$ divided by $3\sqrt{3}$?

▶ Divide the coefficients → $\dfrac{6}{3} = 2$.

▶ Divide the roots → $\dfrac{\sqrt{15}}{\sqrt{3}} = \sqrt{5}$.

▶ Put the two parts together → $2\sqrt{5}$.

What happens if one of those parts does not divide evenly? For the coefficients, it is not a problem. Just write them as a fraction:

$$5\sqrt{2} \div 3\sqrt{2} = \frac{5}{3} \times \frac{\sqrt{2}}{\sqrt{2}} = \frac{5}{3} \times 1 = \frac{5}{3}$$

For the roots, it can be tricky because the conventions of math don't allow us to have a root as the denominator of a fraction. That means we have to take an extra step to convert a denominator with a root into an integer:

$$\sqrt{3} \div \sqrt{2} = \frac{\sqrt{3}}{\sqrt{2}}$$

How do we get the root out of the denominator? Remember that we can multiply any number by 1 without changing the number? We will do that here, but we will write the 1 in a way that helps us, $\dfrac{\sqrt{2}}{\sqrt{2}}$:

$$\frac{\sqrt{3}}{\sqrt{2}} \times \frac{\sqrt{2}}{\sqrt{2}} = \frac{\sqrt{6}}{\sqrt{4}} = \frac{\sqrt{6}}{2}$$

Now the fraction has an integer in the denominator.

▶ What is $7\sqrt{5}$ divided by $3\sqrt{2}$?

▶ Divide the coefficients $\rightarrow \dfrac{7}{3}$.

▶ Divide the roots $\rightarrow \dfrac{\sqrt{5}}{\sqrt{2}}$.

▶ Put the two parts together $\rightarrow \dfrac{7}{3} \times \dfrac{\sqrt{5}}{\sqrt{2}} = \dfrac{7\sqrt{5}}{3\sqrt{2}}$.

▶ Since there is a root in the denominator, we need to change that. We need to multiply $\sqrt{2}$ by $\sqrt{2}$ to change it to an integer, so we need to multiply the whole fraction by 1 in the form of $\dfrac{\sqrt{2}}{\sqrt{2}}$.

$$\dfrac{7\sqrt{5}}{3\sqrt{2}} \times \dfrac{\sqrt{2}}{\sqrt{2}} = \dfrac{7\sqrt{10}}{3\sqrt{4}} = \dfrac{7\sqrt{10}}{3 \times 2} = \dfrac{7\sqrt{10}}{6}$$

Scientific Notation

Scientific notation is a way to use exponents to express very large or very small numbers. It is difficult to write the distance from here to the sun because the number is so large. Instead of writing 92,960,000 miles every time we mention that distance, we can say 9.296×10^7 miles.

Scientific notation is based on powers of 10. To express a number in scientific notation, we write the number in two parts: the digits of the number, not counting any zeros at the end, and 10 to the power of whatever puts the decimal point after the first digit of the number. Let's look at an easy number first: 300. The "digits" part of the number, not counting any zeros at the end, is just a 3. We write the 3 and put a decimal point after it. Then we see how many powers of 10 it takes to put the decimal in that position: $300 = 3 \times 100 = 3 \times 10^2$, so it's two powers of 10. We would write 3.0×10^2.

EXAMPLE

▶ Write 345,900 in scientific notation.

▶ First, we need to write just the digits, without the zeros at the end → 3,459.

▶ Then we put the decimal point after the first digit → 3.459.

▶ Now count how many decimal places we had to move to get the decimal to that position. The original number was 345,900, so the decimal was originally at the end of that → 345,900.0.

▶ Count the places you move to the left to get 3.459: 5 places.

▶ Write the second part of the scientific notation number with 10 to the 5th power → 3.459×10^5.

We also can use scientific notation to express very small numbers. The thickness of a sheet of paper is 0.004 inch. Let's write that in scientific notation.

First, we write just the digit part, without any zeros on the end (in this case, the *front* end): 4. Now we put a decimal point after that digit: 4.0. Now we count how many decimal places we had to move the decimal: 3 places *to the right*:

$$0.004 = 4 \times \frac{1}{1,000}$$

Remember how we did negative exponents? For example, $\frac{1}{1,000}$ equals $\frac{1}{10^3}$, which equals 10^{-3}, and 0.004 in scientific notation is 4×10^{-3}. Just remember that if you are working with a big number and move the decimal to the left, the exponent will be positive. If you are working with a small number and move the decimal to the right, the exponent will be negative.

▶ Write 0.009 in scientific notation.

▶ First, we need to write just the digits, without the zeros at the front → 9.

▶ Then we put the decimal point after the first digit → 9.

▶ Now count how many decimal places we had to move to get the decimal to that position: the original number was 0.009, so we need to move 3 places to the right.

▶ When we move to the right, the exponent will be negative, so we write 10^{-3}: 9×10^{-3}.

Let's look at a few more examples of numbers written in scientific notation so that you get more used to the pattern:

$$730 = 7.3 \times 10^2$$

$$56,00,000 = 5.6 \times 10^6$$

$$3,490,000,000 = 3.49 \times 10^9$$

$$0.03 = 3 \times 10^{-2}$$

$$0.9556 = 9.556 \times 10^{-1}$$

$$0.00574 = 5.74 \times 10^{-3}$$

1.3 equals 1.3×10^1 because it already has the decimal in the proper position!

We can also take a number that is in scientific notation and use our knowledge of powers of 10 to convert it to standard notation. Let's write 4.26×10^4 in standard notation. How many powers of 10 do we have? 4. That means: $4.26 \times 10^4 = 4.26 \times 10,000 = 42,600$.

This process is easier if we just let the exponent on the 10 tell us the number of decimal places to move. This is basically the process of converting to scientific notation in reverse, so positive exponents mean moving to the right to make the number larger, and negative exponents mean moving to the left to make the number smaller. 10^9 means move 9 decimal places to the right, while 10^{-5} means move 5 decimal places to the left.

EXAMPLE

▶ Write 2.3×10^{-2} in standard notation.

▶ 10^{-2} here means move two decimal places to the left → $2.3 \times 10^{-2} = 0.023$.

BTW

Remember that, in scientific notation, a negative exponent on the 10 means the original number is smaller than it may appear and a positive exponent on the 10 means the original number is larger than it may appear.

EXERCISES

EXERCISE 7-1

Write each of the following as exponents.

1. $3 \times 3 \times 3$
2. $5 \times 5 \times 5 \times 5 \times 5 \times 5$
3. 234×234

Write out these exponents as a group of numbers being multiplied.

$4^3 \rightarrow 4 \times 4 \times 4$

4. 78^4

5. 2^6

6. 6^3

EXERCISE 7-2

Multiply each pair of numbers.

1. $2^3 \times 2^5$
2. $54^2 \times 54^9$
3. 6×6^{15}
4. $4^{-3} \times 4^3$

EXERCISE 7-3

Divide each of the following.

1. $\dfrac{7^4}{7^2}$

2. $\dfrac{9^8}{9^8}$

3. $\dfrac{27^{13}}{27^5}$

4. $\dfrac{12^3}{12^6}$

EXERCISE 7-4

Raise each number to the power indicated.

1. $(5^3)^4$

2. $(11^8)^1$

3. $(y^2)^5$

4. $(5x^3y)^4$

EXERCISE 7-5

Write each fraction as a number with a negative exponent.

1. $\dfrac{1}{25}$

2. $\dfrac{1}{81}$

3. $\dfrac{1}{8}$

Write each number as a fraction.

4. 3^{-2}

5. 41^{-7}

6. 1^{-5}

EXERCISE 7-6

Find the square root of each number.

1. $\sqrt{36}$
2. $\sqrt{100}$
3. $\sqrt{49}$

Simplify each number as much as possible.

4. $\sqrt{54}$
5. $\sqrt{44}$
6. $\sqrt{98}$

Estimate the value of each number and write as a decimal.

7. $\sqrt{50}$
8. $\sqrt{110}$
9. $\sqrt{80}$
10. $\sqrt{810}$

EXERCISE 7-7

Add or subtract as indicated in each question.

1. $4\sqrt{5} + 2\sqrt{5}$
2. $\sqrt{5} + 12\sqrt{5}$
3. $6\sqrt{3} - 3\sqrt{3}$
4. $5\sqrt{2} - \sqrt{2}$

EXERCISE 7-8

Multiply each pair of numbers.

1. $\sqrt{2} \times \sqrt{3}$
2. $6\sqrt{2} \times 2\sqrt{5}$

3. $\sqrt{5} \times \sqrt{5}$

4. $2\sqrt{8} \times \sqrt{20}$

EXERCISE 7-9

Divide each of the following as indicated.

1. $\sqrt{6} \div \sqrt{3}$

2. $\sqrt{8} \div \sqrt{2}$

3. $(4\sqrt{5}) \div (2\sqrt{5})$

4. $(12\sqrt{5}) \div (7\sqrt{2})$

EXERCISE 7-10

Write each number using scientific notation.

1. 94,000,000

2. 2,600

3. 0.83

4. 0.0000524

Write each of the following using standard notation.

5. 4.83×10^{7}

6. 9.6×10^{2}

7. 5.462×10^{-4}

8. 8.4×10^{-1}

Flashcard App

8 Equations and Inequalities

MUST KNOW

⚡ An algebraic equation says that two expressions are equal. Solving an equation allows us to find an unknown amount, which we often do in everyday life.

⚡ Anything we do to one side of an equation, we must also do to the other side. If we multiply or divide an inequality by a negative we must also flip the sign around.

⚡ We can solve two equations at the same time if they share the same variables. We call these equations *simultaneous* because we can solve them simultaneously!

I n our daily lives, we are often asked to solve a problem in which one value is unknown. When we add up the prices of two items to see if we have enough money, the total is unknown until we solve for it. In algebra, we use a letter called a **variable** to stand in for the unknown amount. The variables x and y are common, but any letter can be a variable. Since a variable represents a number, we treat it like a number and can perform operations with it. In this chapter, we will learn to write algebraic expressions, equations, and inequalities using variables. We will also learn to solve equations and inequalities.

Algebraic Expressions

An **algebraic expression** is an operation that uses a variable, such as $x + 2$ or $\frac{3}{y}$. When we multiply a variable times a number, we usually do not write the multiplication symbol; instead of $x \times 2$ we would write $2x$.

When we do word problems, we translate the words into math. We read: *If Henri has two cookies and Brie has three cookies, how many cookies do they have together?* To solve the problem, we write $2 + 3 = 5$. The same is true with algebra word problems. We need to be able to read words and write math. As an algebra problem, we could write the cookie question as $2 + 3 = x$. If we are given an equation, we need to be able to translate the math into words we understand in order to solve the equation. Let's look at some common expressions and what they mean:

Expression	Means
$x + 2$	the sum of a number and two
$x - 2$	two less than a number
$2x$	a number times two
$\dfrac{x}{2}$	a number divided by two
$\dfrac{2}{x}$	two divided by a number
$\dfrac{1}{4}x$	one-fourth of a number
x^2	a number squared (times itself)

What if we have a more complicated expression, such as $4x + 14$? How should we write that in words? The first expression is $4x$, which means four times a number. Then we are to add 14 to that result. We might think we should write the words in the order the expressions appear, but what happens if we write *four times a number plus fourteen*? Anyone who read that might think it means $4(x + 14)$. That would give the wrong result!

If you are the only person who will be reading the translation, you can write it in whatever way makes sense to you, as long as you can read it back correctly. If anyone else is going to read your translation, you must be very careful. How can we write $4x + 14$ in a way that cannot be misunderstood? What if we say *fourteen plus a number times four*? That might mean $(14 + x) \times 4$. Ugh. That's not right either. We know we need $4x$, and then we need to add 14. Let's say *the product of four and a number, added to fourteen,* or *fourteen, plus the product of four and a number.*

▶ Write the expression $\dfrac{x-3}{12}$ in words.

▶ This is a division problem. The expression in the numerator can be written as *a number minus three* or *three less than a number*. When we try to include the *divided-by-twelve* part, we need to be careful to avoid confusion. Let's say *a number minus three, then divided by twelve*.

We also, of course, need to be able to translate a word problem into a math problem we can solve. We might read *Jacques has six fewer pairs of shoes than does Rachel*. Since we do not know how many pairs of shoes either of them actually has, we will need variables. To keep it simple, let's use *j* for the number of pairs of shoes Jacques has and *r* for the number of pairs of shoes Rachel has. We can write that expression as $j = r - 6$.

▶ Write the following as an algebraic expression: to buy a toy, Kaveri needs three times as much as her weekly allowance, plus another two dollars.

▶ Since we don't know the amount of Kaveri's allowance, we will need a variable. Let's use *a* for allowance. We also don't know the cost of the toy, so we can use *t* for that.

▶ The toy costs three times her allowance, or $3a$, plus 2 more dollars. We can, therefore, write $t = 3a + 2$.

Equations

An **equation** is a statement that says two expressions are equal. It says what is on the left side of the equal sign is equal to what is on the right side of the equal sign. For example, $x + 4 = 5$ is an equation in which both sides must equal 5.

We can perform operations on equations to solve for an unknown variable, as long as we remember that whatever we do to one side of the equation, we must also do to the other side. If we add 2 to the left side, we must add 2 to the right side. If we divide the right side by 5, we must also divide the left side by 5. Keep it equal!

Solving Equations by Addition and Subtraction

An equation that involves addition or subtraction can be solved by addition or subtraction. The goal in solving any equation is to isolate the variable (get it by itself) on one side of the equals sign with a number on the other, like this: $y = 14$. That tells you the value of the variable. Keep that goal in mind at all times and let it guide you to what you need to do to achieve the goal.

If we are solving for y in the equation $y - 3 = 6$, let's think about how we can get y by itself. Right now, we have $y - 3$ on the left side. We want only y to be there, so let's add 3 to balance out the -3 we have there. Remember, if we add 3 to the left side, we must also add 3 to the right side.

$$y - 3 = 6$$
$$y - 3 + 3 = 6 + 3$$
$$y = 9$$

That's it. We have solved for y!

▶ Solve for x: $12 = x + 7$.

▶ Here, we have the variable on the right side: $x + 7$. Since we have 7 added to x, we will need to subtract the 7 from x to get x by itself. Remember to do that to both sides of the equation.

$$12 = x + 7$$
$$12 - 7 = x + 7 - 7$$
$$5 = x$$

Solving Equations by Multiplication and Division

An equation that involves multiplication or division can be solved by multiplication or division. Remember that the goal in solving any equation is to isolate the variable and that what you do to one side of the equation you must also do to the other side. If we have the equation $3x = 12$, the expression $3x$ involves multiplication. To "undo" the multiplication by 3, we need to *divide* by 3, and we must do that to both sides:

$$3x = 12$$
$$\frac{3x}{3} = \frac{12}{3}$$
$$x = 4$$

▶ Solve for y: $20 = 5y$.

▶ Here, we have the variable on the right side: $5y$. Since we have 5 multiplied by y, we need to divide by 5 to undo that multiplication. Remember to do that to both sides of the equation.

$$20 = 5y$$
$$\frac{20}{5} = \frac{5y}{5}$$
$$4 = y$$

Equations that involve division work in much the same way. Since multiplication is the opposite of division, we will need to multiply to undo the division. If we want to solve $\frac{x}{4} = 2$, we will need to figure out what to do to the equation to isolate x. Since the expression is x divided by 4, we can multiply by 4 to undo the division:

$$\frac{x}{4} \times 4 = 2 \times 4$$
$$\frac{x}{4} \times \frac{4}{1} = 2 \times 4$$
$$\frac{4x}{4} = 8$$
$$x = 8$$

What if the expression were $\frac{4}{x}$ instead of $\frac{x}{4}$? Multiplying by 4 would not get x by itself. Let's see how that equation looks: $\frac{4}{x} = 2$.

Our goal is to isolate x, but x is in a fraction. We have to get it out of the fraction, and the easiest way to do that will be to multiply the fraction by whatever is in the denominator, as we did before. In this case, instead of there being a 4 in the denominator, there is an x. Does that matter? No. Multiply both sides of the equation by x and see what happens.

$$\frac{4}{x} = 2$$

$$\frac{4}{x} \times x = 2 \times x$$

$$\frac{4}{x} \times \frac{x}{1} = 2x$$

$$\frac{4x}{x} = 2x$$

$$4 = 2x$$

Now we have the x out of the fraction and can simply divide both sides by 2 to isolate x:

$$4 = 2x$$

$$\frac{4}{2} = \frac{2x}{2}$$

$$2 = x$$

EXAMPLE

▶ If $\dfrac{10}{y}$ equals 5, what is the value of y?

▶ Multiply both sides by y to get the variable out of the fraction.

$$\frac{10}{y} \times y = 5 \times y$$

$$10 = 5y$$

▶ Now divide both sides by 5 to isolate y.

$$10 = 5y$$
$$\frac{10}{5} = \frac{5y}{5}$$
$$2 = y$$

▶ For those last two problems, we did two steps to solve them. It is common for equations to require multiple steps, so just take it slow and do one thing at a time. Keep the goal of isolating the variable in mind and keep moving toward that goal.

EXAMPLE

▶ Solve for k: $42 = 12 + 6k$.

▶ This problem involves both addition and multiplication, so we should expect to have to do subtraction and division in order to solve it. Let's do the subtraction first to remove the 12 that is added to $6k$.

$$42 = 12 + 6k$$
$$42 - 12 = 12 + 6k - 12$$
$$30 = 6k$$

▶ Now we need to undo the multiplication by dividing both sides by 6.

$$\frac{30}{6} = \frac{6k}{6}$$
$$5 = k$$

All we have to do is take things step by step. Let's try another with a different twist.

EXAMPLE

▶ Solve for b: $2b + 1 = b + 3$.

▶ This equation has the variable b on both sides of the equation. We need to get that variable on only one side and a number on the other side. Let's start by gathering all the bs together. If we subtract a b from each side, that will totally remove the b from the right side.

$$2b + 1 = b + 3$$
$$2b + 1 - b = b + 3 - b$$
$$b + 1 = 3$$

▶ Now we have b on only one side, but it is $b + 1$, so we will need to subtract 1 to get b by itself. Do that to both sides.

$$b + 1 = 3$$
$$b + 1 - 1 = 3 - 1$$
$$b = 2$$

▶ If you want to check your work, put 2 in for b in the original equation and see if it works.

$$2b + 1 = b + 3$$
$$2(2) + 1 = 2 + 3$$
$$4 + 1 = 5$$
$$5 = 5$$

▶ The solution works, so we know we did it correctly. You can always check your work this way to be sure of your answer.

Some equations have more than one solution. Look at $x^2 = 9$. The value of x could be either 3 or -3. Some equations even have an infinite number of solutions! It may not be easy to tell that until you have started working to solve the equation, but eventually you will see that any number you choose will solve the equation.

EXAMPLE

▶ Solve for p: $3p + 8 - 3p = 8$.

▶ First, let's combine the variables on the left side.

$$3p - 3p = 0$$

▶ That leaves us with $8 = 8$, which is pretty obvious, right? Any number we use for p will work. Try a few. Let $p = 3$.

$$3p + 8 - 3p = 8$$
$$3(3) + 8 - 3(3) = 8$$
$$9 + 8 - 9 = 8$$
$$8 = 8$$

▶ Let $p = 0$.

$$3p + 8 - 3p = 8$$
$$3(0) + 8 - 3(0) = 8$$
$$0 + 8 - 0 = 8$$
$$8 = 8$$

▶ Let $p = -2$.

$$3p + 8 - 3p = 8$$
$$3(-2) + 8 - 3(-2) = 8$$
$$-6 + 8 - (-6) = 8$$
$$-6 + 8 + 6 = 8$$
$$8 = 8$$

▶ Yup. Always works. This equation has infinite solutions.

Some equations have no solution. Again, it may not be easy to tell that until you have started working to solve the equation, but eventually you will see that nothing you try will work.

Solve for w: $7w + 14 - 3w = 92 + 4w$.

First, let's get the variables all on the same side. Combine $7w - 3w$.

$$4w + 14 = 92 + 4w$$

Subtract $4w$ from both sides.

$$4w + 14 = 92 + 4w$$
$$14 = 92$$

Um ... that is obviously *not* true. That's how we know that this equation has no solution.

Inequalities

An equation is a mathematical expression that says two quantities are equal. An **inequality** also relates two expressions but says that they are *not* necessarily equal. Here are the symbols we use for inequalities:

>	greater than
<	less than
≥	greater than or equal to
≤	less than or equal to

When we see a statement such as $y > -4$, we read it as *y is greater than negative 4*. On a number line, that looks like this:

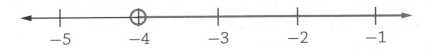

We use an open circle dot to show that −4 itself is *not* a solution but that everything greater than –4 is.

When we see a statement such as $x \leq 3$, we read it as *x is less than or equal to 3*. On a number line, that looks like this:

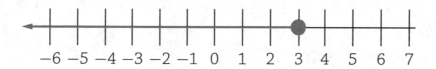

We use a solid dot to show that 3 *is* a solution, along with anything less than 3.

Solving Inequalities by Addition and Subtraction

Solving an inequality is very similar to solving an equation. To solve an inequality involving addition or subtraction, we will use addition or subtraction, and the process for solving is exactly the same. We can add or subtract things from both sides of the inequality just as we did with equations. The goal is the same, too: isolate the variable on one side of the sign and leave just a number on the other side.

Let's try one: $x + 3 > 7$.

Since this inequality has 3 added to x, we need to subtract 3 from both sides:

$$x + 3 - 3 > 7 - 3$$
$$x > 4$$

That's all there is to it. We just keep the inequality sign the same and follow the same steps we did to solve equations by adding and subtracting.

BTW

We deal with inequalities often in our lives, though we may not think of them that way. When I drive on the highway, there are speed limit signs. If the speed limit is 55 miles per hour, that doesn't mean I have to drive exactly 55 miles per hour; it just means I can't (legally) go over 55 miles per hour. In math terms, the allowed speed is $s \leq 55$.

▶ Simplify the following inequality:

$$2q - 4 \leq 20 + q$$

▶ First, let's get the variables all on the same side. Subtract q from both sides.

$$2q - 4 - q \leq 20 + q - q$$
$$q - 4 \leq 20$$

▶ Now add 4 to both sides.

$$q - 4 + 4 \leq 20 + 4$$
$$q \leq 24$$

Solving Inequalities by Multiplication and Division

Solving an inequality involving multiplication or division is almost the same as solving an equation, but there's one important twist: If you multiply or divide an inequality by a negative number, you must flip the inequality sign around to point in the opposite direction. Why? Think about it this way: $3 > 2$ but if we multiply both sides by -1, we get $-3 > -2$. That doesn't work on the number line, so multiplying by a negative number reverses the direction of the sign. The same is true for division.

If you multiply or divide by a positive number, it's exactly the same as solving an equation by multiplying or dividing. If we want to solve $3x \geq 12$, all we need to do is divide both sides by 3: $\dfrac{3x}{3} \geq \dfrac{12}{3}$, so $x \geq 4$.

EXAMPLE

▶ Let's simplify this inequality: $\dfrac{8}{y} > 4$.

▶ First, let's get y out of the fraction. Multiply both sides by y.

$$\dfrac{8}{y} \times y > 4 \times y$$

$$8 > 4y$$

▶ Now divide both sides by 4 to isolate y.

$$\dfrac{8}{4} > \dfrac{4y}{4}$$

$$2 > y$$

▶ It is traditional to write the solutions to equations and inequalities with the variable on the left side, like this: $x = 4$. To do that with an inequality, be careful that you keep the inequality sign pointing in the correct direction.

▶ The solution to our example is $2 > y$, which means *2 is greater than y*. If we write it with y on the left side, we need to write $y < 2$, which keeps the meaning that *2 is greater than y*. In other words, if you rewrite an inequality, keep the sign pointing at the same value it was originally pointing at.

Just as with solving equations, solving inequalities often requires more than one step. We may need to add or subtract *and* multiply or divide.

▶ Solve for y: $7y + 5 < 8 + 2y$.

▶ First, let's get both y terms on the same side of the inequality. Subtract $2y$ from both sides.

$$7y + 5 - 2y < 8 + 2y - 2y$$
$$5y + 5 < 8$$

▶ Now let's get all the numbers on the other side of the inequality. Subtract 5 from both sides.

$$5y + 5 - 5 < 8 - 5$$
$$5y < 3$$

▶ Now we can isolate y. Since y is multiplied by 5, we need to divide both sides by 5.

$$5y < 3$$
$$\frac{5y}{5} < \frac{3}{5}$$
$$y < \frac{3}{5}$$

Let's try one more, this time with a negative number.

▶ Solve for f: $-6f > -14 + f$.

▶ First, let's get all the fs on the same side of the inequality. Subtract f from both sides.

$$-6f - f > -14 + f - f$$
$$-7f > -14$$

▶ Now, we need to isolate f. On the left side of the inequality, f is multiplied by -7, so to undo that multiplication, we need to divide by -7.

$$\frac{-7f}{-7} > \frac{-14}{-7}$$

▶ Since we divided by a negative number, we must flip the sign.

$$\frac{-7f}{-7} < \frac{-14}{-7}$$

$$f < 2$$

BTW

Remember, if you multiply or divide an inequality by a negative number, flip the sign!

Simultaneous Equations

We can even solve equations that have more than one variable, if we have more than one equation. We call these **simultaneous equations**. Let's look at a pair of equations that both contain the variables x and y:

$$2x + y = 11$$
$$x + y = 7$$

If we want to solve for the value of x, we can do that in a couple of ways:

■ **Substitution**   We use one of the equations to solve for one variable in terms of the other and then substitute that value into the second equation. We usually choose the simplest equation to solve for a variable. For this pair, I would choose the second equation.

$$2x + y = 11$$
$$x + y = 7$$

We can take $x + y = 7$ and solve the equation for x in terms of y. Subtract y from both sides to isolate x: $x + y - y = 7 - y$, so $x = 7 - y$. Now we substitute that into the other equation: $2x + y = 11$ becomes

$2(7 - y) + y = 11$. Now we have one equation with one variable, and we can solve it.

$$2(7 - y) + y = 11$$

Distribute the 2 across the parentheses:

$$2(7) - (2)(y) + y = 11$$
$$14 - 2y + y = 11$$

Combine the y terms:

$$14 - y = 11$$

Subtract 14 from both sides to isolate y.

$$14 - y - 14 = 11 - 14$$
$$-y = -3$$

Multiply both sides by -1 to solve for y:

$$-y(-1) = -3(-1)$$
$$y = 3$$

Now we have the value of y, so we can plug that into either equation to solve it for x. Again, I choose the simpler equation: $x + y = 7$.

$$x + 3 = 7$$
$$x = 4$$

- **Elimination**   We manipulate the equations in such a way as to eliminate one of the variables. We do this by adding or subtracting one equation from the other:

$$2x + y = 11$$
$$x + y = 7$$

For this pair of equations, if we subtract the bottom equation from the top equation, the variable y will drop out.

$$2x + y = 11$$
$$-(x + y = 7)$$

Make sure you distribute the minus sign across everything in parentheses:

$$2x + y = 11$$
$$-x - y = -7$$

Now we just combine the terms straight down:

$$
\begin{array}{rrcr}
2x & +y & = & 11 \\
-x & -y & = & -7 \\
\hline
x & 0 & = & 4
\end{array}
$$

Now we have $x = 4$. We can plug that into either equation to find the value of y.

$$x + y = 7$$
$$4 + y = 7$$
$$y = 3$$

We can always test our solutions in both equations to be sure we have done everything correctly.

$$2x + y = 11$$
$$2(4) + 3 = 11$$
$$11 = 11$$

That works. Now:

$$x + y = 7$$
$$4 + 3 = 7$$
$$7 = 7$$

That works too, so our solutions are correct.

▶ Solve the following system of equations for the value of x:

$$7x + y = 12$$
$$16y = -32$$

▶ This set is a good candidate for the substitution method because it will be easy to solve the second equation for y, and then we can plug that into the first equation to solve for x.

$$16y = -32$$

▶ Divide both sides by 16:

$$\frac{16y}{16} = \frac{-32}{16}$$

▶ Now we have $y = -2$ and we can use that value of y in the first equation:

$$7x + y = 12$$
$$7x + (-2) = 12$$

▶ Add 2 to both sides:

$$7x + (-2) + 2 = 12 + 2$$

▶ Now we have $7x = 14$. Divide both sides by 7 to solve for x.

$$\frac{7x}{7} = \frac{14}{7}$$
$$x = 2$$

Let's try one with the elimination method.

▶ Solve the following system of equations for the value of x:

$$23x - 2y = 34$$
$$3x + 2y = 18$$

▶ This set is a good candidate for the elimination method because the y terms will be eliminated if we add the equations together.

$$23x - 2y = 34$$
$$+(3x + 2y = 18)$$
$$26x = 52$$
$$x = 2$$

▶ The question asked us to find the value of x, which we have done. If we needed to also solve for y, we could use the value we found for x to find the value of y. The second equation looks simpler, so let's use that one: $3x + 2y = 18$ becomes $3(2) + 2y = 18$.

$$6 + 2y = 18$$
$$2y = 12$$
$$y = 6$$

Sometimes, we may need to manipulate one of the equations to make it do what we need it to do in order to eliminate a variable. We do this by multiplying both sides of the equation by some number. Let's look at this pair of equations and solve them for the value of y.

$$3x + 2y = 22$$
$$2x + 3y = 23$$

Substitution would be difficult here, so elimination is the way to go. Elimination won't happen by simply adding or subtracting the equations as they are. What we need to do is get one of the variables to drop out. We are looking for the value of y, so we want x to drop out. How can we do that? If we multiply the top equation by 2 and the bottom equation by -3, we can make the x terms cancel each other out. Remember that what we do to one side of an equation we must also do to the other side.

$$2(3x + 2y) = 2(22)$$
$$-3(2x + 3y) = -3(23)$$

Now our pair of equations looks like this:

$$6x + 4y = 44$$
$$-6x - 9y = -69$$

When we add these up, the x terms cancel out, leaving us with $-5y = -25$. Divide both sides by -5 to solve for y: $y = 5$.

Now, let's work one out together.

EXAMPLE

▶ Solve the following system of equations for the values of x and y.

$$11x + 8y = 14$$
$$5x + 2y = 8$$

▶ Look at both equations and see what we can do to eliminate a variable. The x terms would have to be multiplied by a fraction, so let's avoid that if we can. We can eliminate the y terms if we multiply the second equation by -4.

$$11x + 8y = 14$$
$$-4(5x + 2y) = -4(8)$$

▶ Now our pair of equations looks like this:

$$11x + 8y = 14$$
$$-20x - 8y = -32$$

▶ When we add these up, the y terms cancel out, leaving us with $-9x = -18$.

▶ Divide both sides by -9 to solve for x: $x = 2$.

▶ Since we need to find both x and y, we can use 2 for x in one of the equations to solve for y. The second equation looks a bit simpler, so let's use that one.

$$5x + 2y = 8$$
$$5(2) + 2y = 8$$
$$10 + 2y = 8$$
$$2y = -2$$
$$y = -1$$

▶ For this pair of equations, $x = 2$ and $y = -1$.

We can also solve simultaneous equations by graphing. Let's look at our first pair:

$$2x + y = 11$$
$$x + y = 7$$

Each equation can be graphed as a line. We can make a table of possible values for x to see what the corresponding y values are.

For the first equation: $2x + y = 11$.

x	y
−1	13
1	9
3	5

Now we can plot each of these pairs of coordinates on the coordinate plane and draw the line.

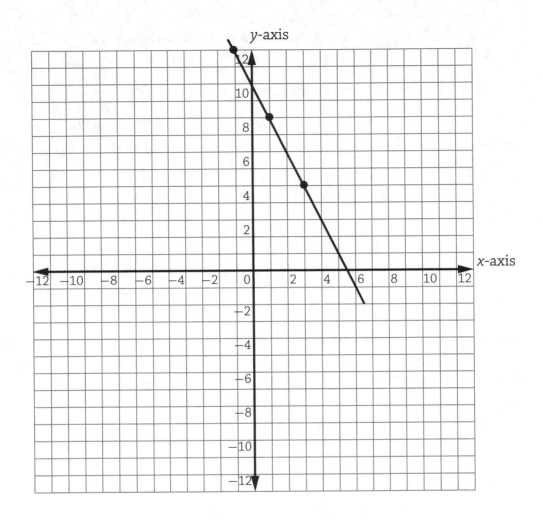

For the second equation: $x + y = 7$.

x	y
−1	8
1	6
3	4

Now we can plot each of these pairs of coordinates on the coordinate plane. The place where the lines intersect shows us the values for x and y. In this case, the lines intersect at (4, 3), which means $x = 4$ and $y = 3$.

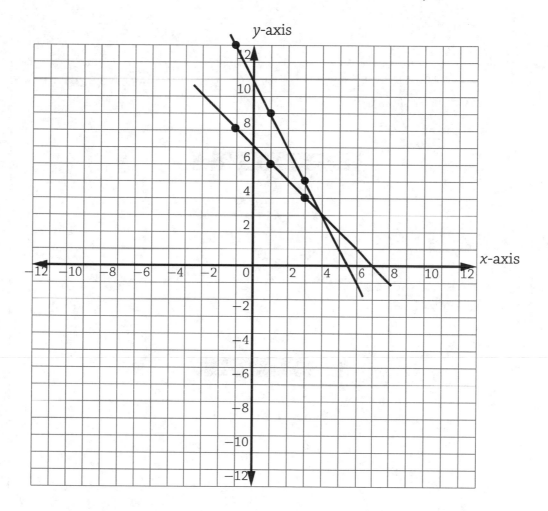

▶ Solve this system of equations by graphing:

$$\frac{1}{2}x - y = 2$$
$$2x + y = 3$$

▶ For $\frac{1}{2}x - y = 2$:

x	y
−2	−3
2	−1
4	0

▶ For $2x + y = 3$:

x	y
−2	7
2	−1
4	−5

▶ Graph both lines.

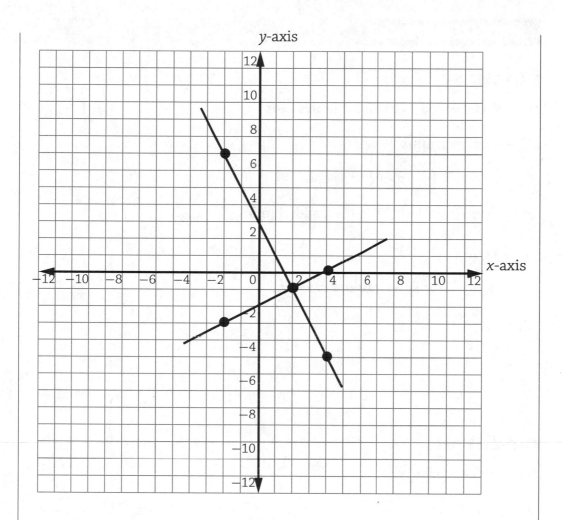

▶ The intersection of the lines on the graph is at (2, −1), which means
$x = 2$ and $y = −1$.

EXERCISES

EXERCISE 8-1

Write the following as algebraic expressions.

1. a number minus eleven

2. two times the product of a and b

3. a number divided by 7, then added to 5

EXERCISE 8-2

Write these algebraic expressions in words.

1. $y + 106$

2. $\dfrac{12}{x}$

3. $4x - 3.5$

EXERCISE 8-3

Solve each equation for x.

1. $x - 8 = 12$

2. $5 + x = 7$

3. $\dfrac{x}{3} = 4$

4. $7x = 49$

5. $\dfrac{10}{x} = 2$

6. $4x + 6 = 38$

7. $3(x + 2) = 3x + 11$

8. $\dfrac{2}{x} + 5 = 6$

9. $x + y = 14$

10. $4(2 + x) = 4x + 8$

EXERCISE 8-4

Solve each inequality for x.

1. $x + 4 < 7$

2. $-2 + x > 9$

3. $\dfrac{x}{5} \le 4$

4. $12x \ge 40$

5. $\dfrac{8}{x} < 1$

6. $3x - 4 > 86$

7. $\dfrac{20}{x} + 3 \ge 7$

8. $-2x - 4 \le 3x + 6$

EXERCISE 8-5

Questions 1–3 refer to the following system of equations:

$$x + y = 11$$
$$3x - y = 9$$

1. Solve for x and y by substitution.

2. Solve for x and y by elimination.

3. Solve for x and y by graphing.

Flashcard App

Data Presentation

MUST ⚡ KNOW

⚡ In order to better understand data that has been collected or to present that data to others, we use various types of charts and graphs.

⚡ Bar graphs show values that are independent of one another.

⚡ Line graphs show how a value or values change over time.

⚡ Stem-and-leaf plots, box-and-whisker plots, and histograms show how data is distributed. They are useful for finding averages, medians, and ranges of values.

⚡ Circle graphs and Venn diagrams show proportions and are useful for showing results of surveys.

Data is a group of facts. It may be numbers, measurements, or descriptions of things. If we ask students what flavor ice cream they prefer, their answers are data. When nurses record the heights of patients, those measurements are data. When a grocery store lists the prices of its fresh fruits, the prices are data. There are many, many ways to present data. Lists, charts, graphs, and diagrams are just a few of the ways to represent data. In this chapter, we will look at some of the most common ways to present data and how to interpret graphical representations of data.

BTW

Whenever you are presented with data in any form, it is important that you read all the labels and descriptions given. If there is a title, read it. Read any data labels that are given so that you know what is being measured or compared and what units the data are in.

Bar Graphs

A **bar graph** uses solid bars to show two types of measurement for data in different categories. They are also sometimes called **bar charts** or **bar diagrams**. Bar graphs are common for showing distance over time or the rate at which people do a particular task.

The bars on a bar graph are most commonly shown vertically, like this:

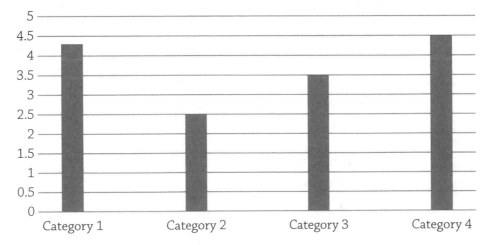

Vertical Bar Graph

A vertical bar graph may also be called a **column graph**, because the bars look like columns.

Bar graphs can also be shown horizontally:

Horizontal bar graph

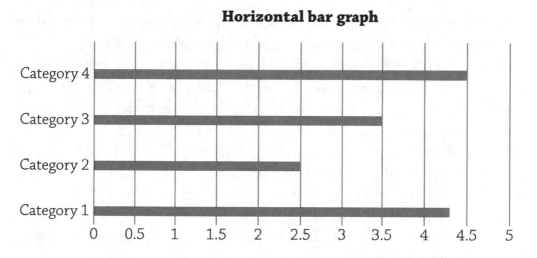

The most important thing to know about a bar graph is that the *longer the bar for a category is*, the greater the value it represents. For both the charts shown above, category 4 has the greatest value, and category 2 has the smallest value.

Let's look at all the parts of a bar graph:

Ron's Marathon Training

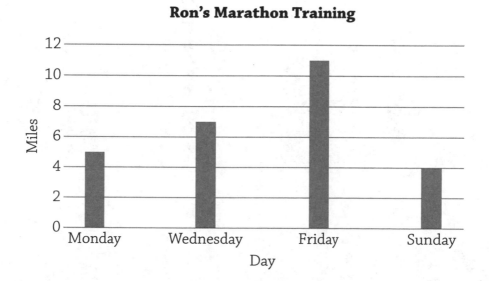

The title of this graph is "Ron's Marathon Training." The numbers along the left side of the graph represent the number of miles Ron runs. The numbers increase by two for each line on the graph. The labels along the bottom of the graph show the days Ron ran. We can assume from the graph that he did not run on Tuesday, Thursday, or Saturday because those days are not labeled on the graph, and there is no data between the days that are shown.

This is a vertical bar graph, and the bars represent the number of miles Ron ran on various days. We can follow a bar for a particular day to find out how far he ran on that day. On Monday, the bar stops halfway between 4 and 6 miles, so we can estimate that he ran 5 miles on Monday. On Wednesday, he ran about 7 miles. On Friday, he ran about 11 miles, and, on Sunday, he ran 4 miles. The bar graph format allows us to easily see that he ran the most miles on Friday and the least on Sunday.

Bar graphs may also show data for more than one category. For example, a chart like the one we just examined might show data for more than one person, such as Ron and his training partner. In that case, the training partner would also have bars shown, usually in a different color. There would be a legend to show which bar goes with which person. It might look like this:

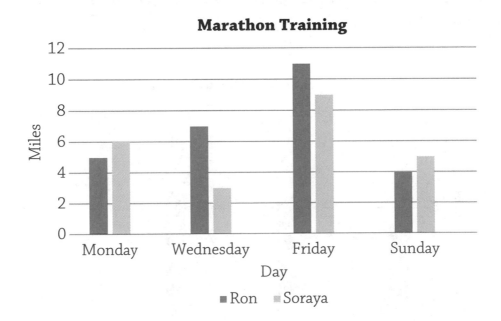

The legend is below the chart and tells us that the dark bars represent Ron's miles and the light bars represent Soraya's miles. We can see from the graph that Soraya ran 6 miles on Monday, about 3 miles on Wednesday, about 9 miles on Friday, and about 5 miles on Sunday. Having two sets of data on the same chart allows us to compare them easily. We can see that Soraya ran more miles than did Ron on Monday and Sunday. We can see that Ron ran more than twice as many miles as Soraya on Wednesday.

EXAMPLE

▶ According to the below bar graph, how many notebooks are needed for Mrs. Babin's class? How many for Mr. Harive's? Which type of school supply requires the same number for both classes?

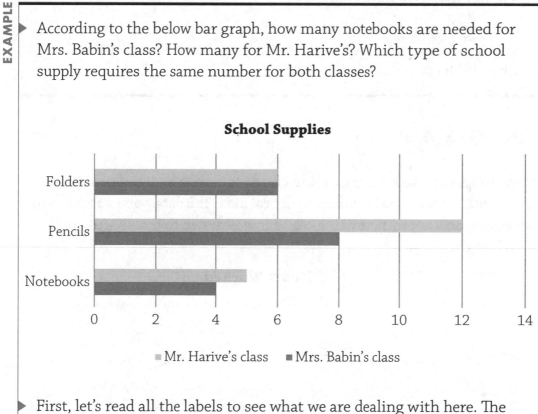

First, let's read all the labels to see what we are dealing with here. The chart is called "School Supplies," and the labels on the left side are types of school supplies. There are numbers along the bottom, which probably represent the number of each type of supply. A legend at the bottom tells us that the light bar represents Mr. Harive's class and the dark bar represents Mrs. Babin's class.

▶ To find out how many notebooks are needed in each class, we look at the left side to see that notebooks are the bottom category, and then we follow those bars across to see where they stop. Four notebooks are needed for Mrs. Babin's class (the dark bar), and five are needed for Mr. Harive's class (the light bar).

▶ To see which supply has the same amount needed for both classes, all we have to do is see which category has bars that are even with each other. Both classes need six folders.

Line Graphs

A **line graph** is also known as a **line chart**. It looks a lot like a coordinate plane, with a horizontal axis and a vertical axis. The data points are plotted like points on a coordinate plane, and a line is drawn to connect them. It looks like this:

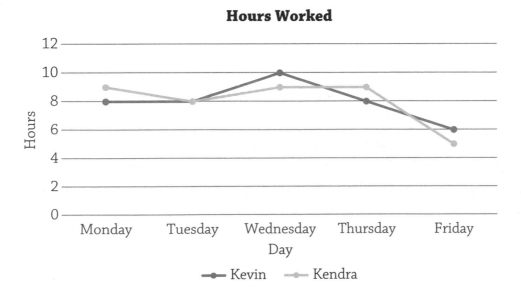

Hours Worked

A line graph makes it easy to see how values change over time. If more than one line is plotted, as with this graph, it allows us to compare the lines at any point and over time. This graph shows us the number of hours worked by Kevin and Kendra on each of the five days of the week. We can see things such as that Kevin worked longer on Wednesday than on any other day, both worked the fewest hours on Friday, Kevin and Kendra worked the same number of hours on Tuesday, et cetera.

EXAMPLE

▶ The below graph shows the monthly average high temperature in Chicago. According to the graph, which month has the second lowest average high temperature?

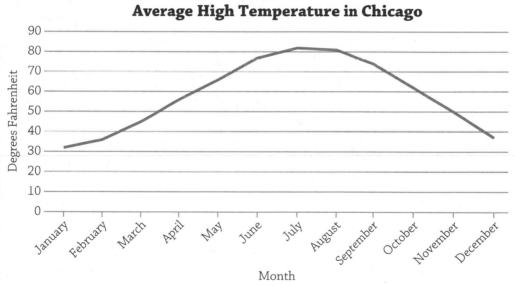

Average High Temperature in Chicago

▶ A line graph allows us to quickly track the temperature over time. Here, we need to focus on the lowest temperature.

▶ January has the lowest temperature at about 32 degrees, and February has the second lowest at about 36 degrees.

The line graphs we have seen so far both show **discrete** data. This means that each data point is plotted as a dot on the graph. We can then draw a line to connect the dots to show trends, or we can leave the dots without a line. Line graphs are also useful for showing **continuous** data—data that is always being measured, such as miles per hour or distance walked. When a graph shows continuous data, the line is smooth and does not show particular points marked. Any spot on the line can be read for data.

▶ The following graph shows the speed of three cars in a race. According to the graph, which car was ahead after 1.5 hours?

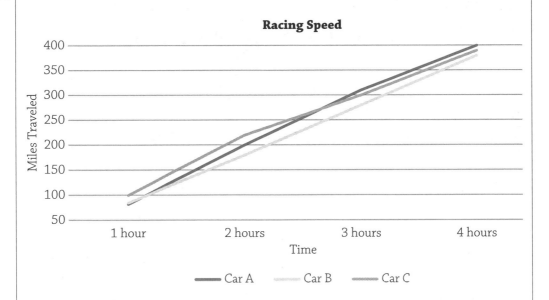

▶ The graph is marked in 1-hour intervals but, since it is a continuous line graph, we can look between the marks for 1 hour and 2 hours to see the distance traveled after 1.5 hours.

▶ Follow that spot up to see where each car is. Car *C* is at about 170 miles, while car *A* is at about 150 miles, and car *B* is at about 135 miles. Car *C* is ahead after 1.5 hours.

Circle Graphs

A **circle graph** is also known as a **pie chart**. It is a circle with each category marked as a sector of the circle—like a wedge of a pie. It shows categories of data with percentages that add up to 100%. It is important to note that a circle graph may not show the actual values, since it only shows percentages. (Twenty percent of 386 is not the same as 20% of 19.) In order to know what the actual values are, we need the actual total. Then we can calculate the value of a certain percent of that total. The best thing about a circle graph is that it allows us to quickly see what parts make up something and what the relative importance of each part is.

BTW

Bar graphs and line graphs show two different types of data, such as distance and time. They can compare data from more than one source, such as the walking speed for three different people. Circle graphs, stem-and-leaf plots, and box-and-whisker plots show how a single set of data is distributed.

Favorite Ice Cream Flavor

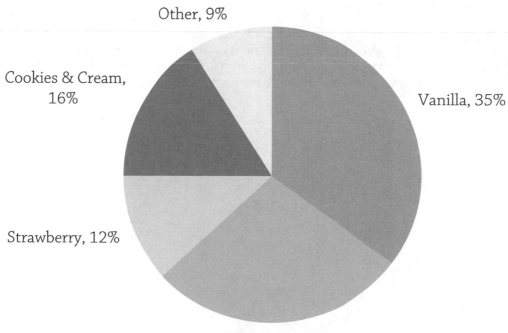

Other, 9%

Cookies & Cream, 16%

Vanilla, 35%

Strawberry, 12%

Chocolate, 28%

This is a graph of the favorite ice cream flavors of students at a local school. Each colored wedge of the circle shows a different flavor and the percentage of students that voted for that flavor. The pie graph format allows us to easily see that vanilla is the most popular flavor, with chocolate not far behind.

Since the circle graph shows only the percent of the total for each flavor, we need to know how many students were surveyed to say how many voted for each flavor. It there were 200 students, and 12% voted for strawberry, then we can take 12% of 200 to find that 24 students voted for strawberry. If there were 50 students, we would take 12% of 50 to find that 6 students voted for strawberry.

Circle graphs often show a category for *Other*. This category counts everything that is not included in the labeled categories. For the ice cream flavor chart, the *Other* category adds up to 9%. This category might include chocolate chip, butter pecan, Neapolitan, mint, peanut butter, and cookie dough. The percent that voted for each of those flavors may be too small to really show on a circle graph, so the designer of the chart groups them all together as *Other*. Some graphs may list what is contained in the *Other* category, but many do not.

 IRL Circle graphs are very useful for financial applications. Companies use them to show sales, industries use them to show the market share of various companies, and individuals can use them for their household budgets.

EXAMPLE

▶ The next chart shows the blood type distribution for a group of 300 people. According to the graph, how many people have blood type O?

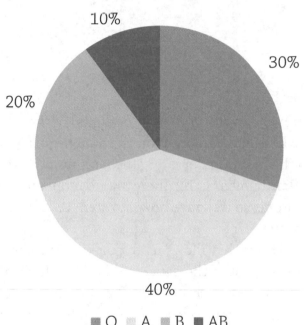

Blood Type

10%

30%

20%

40%

■ O   ■ A   ■ B   ■ AB

▶ Since the chart represents the blood type distribution of 300 people, we will need to find what percent of that group has type O.

▶ According to the chart, 30% have type O, so we need to find 30% of 300: $\dfrac{30}{100} \times 300 = 30 \times 3 = 90$.

Stem-and-Leaf Plots

A **stem-and-leaf plot** is a type of table that allows us to see how numerical data is distributed. It looks like this:

Stem	Leaf
1	3, 4
2	1, 5, 7, 7
3	2
4	

The stem column shows the first digit (or digits) of the data, and the leaf column shows the last digit. On our plot, the data begins with the stem 1 and the leaf 3, so the first number is 13. The next number also has stem 1, but the leaf is 4. That number is 14. There are no more numbers that start with a tens digit of one, so we move down to stem 2. The numbers there are 21, 25, 27, and another 27.

Looking at a chart like this allows us to quickly see where the bulk of the data lies. For the plot, most of the data is in the 20s. There is no data past 32. Stem-and-leaf plots are most useful when the amount of data is fairly small, the data are concentrated (fairly close together), and the number values are small. They also have the advantage of presenting the data in increasing order.

EXAMPLE

▶ The chart here shows the ages of the participants in a ballroom dancing class. What is the range of the ages? What is the most common age?

Stem	Leaf
1	2, 5, 6
2	1, 5, 7, 7, 8
3	2, 2, 4, 6, 8, 9, 9, 9
4	4, 8, 9
5	3, 6, 6
6	1, 7
7	0
8	
9	

▶ The smallest age is 12, and the largest is 70, so the range is: $70 - 12 = 58$ years. Having the data listed in order in a stem-and-leaf plot makes finding the range of data easy.

▶ It is also easy to find the most common data point (the mode). Look for a number that repeats: 27 appears twice, as does 32. The age 39 appears three times, and 56 appears twice. The most common age is 39.

Box-and-Whisker Plots

A **box-and-whisker plot** may also be called a **box plot**. It is another type of chart that allows us to see how numerical data is distributed. It looks like this:

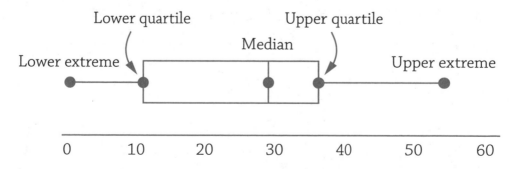

A box-and-whisker plot divides the data into four equal parts called the first, second, third, and fourth quartiles. A **quartile** is just a fancy word for quarter, or one-fourth of the data. A box is drawn around the second and third quartiles in order to highlight the middle half of the data. A vertical line cuts through the box at the median (middle value). There are also markers for the smallest and largest numbers for the data set. These are called the **lower** and **upper extremes**.

Box-and-whisker plots can be drawn horizontally or vertically. Since the one here is drawn horizontally, let's draw a vertical box-and-whisker plot for a set of data.

Our data set is the math test scores for a class: {72, 66, 84, 95, 78, 82, 80, 83, 86, 79, and 90}. The first thing we need to do is put the data in increasing order: {66, 72, 78, 79, 80, 82, 83, 84, 86, 90, and 95}. Now let's identify the median, the middle value. Count how many values there are. There are 11, so the middle value will be the sixth number: 82. When we draw the chart, we go from the lower extreme (66) to the upper extreme (95). We need to draw a box around the middle half of the data, and to do that, we need to know the lower quartile and the upper quartile. To find these values, take the median of the group of numbers below the median for the lower quartile and the median of the group of numbers above the median for the upper quartile:

Median: 82
Lower quartile: 78 (the median of 66, 72, 78, 79, and 80)
Upper quartile: 86 (the median of 83, 84, 86, 90, and 95)

We are ready to draw. The upper extreme is 95, the lower extreme is 66, the box goes from 78 to 86, and the line for the median is at 82.

Math test scores

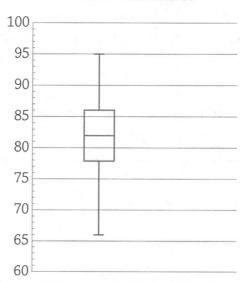

This chart allows us to see that the majority of students scored between 78 and 86 (inside the box). We can also see the median score, the highest score, and the lowest score very easily.

EXAMPLE

▶ The below chart shows race times for a group of runners. What was the range of times for the fastest quartile?

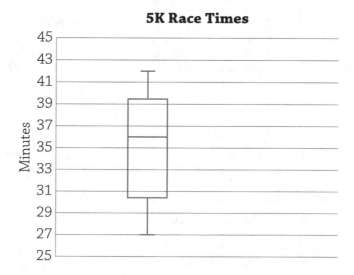

▶ To answer this question, we need to find the fastest quartile. Times along the left side go from 27 minutes up to 42 minutes, so the fastest quartile will be what we normally call the **lower quartile** of the box-and-whisker plot. The lower quartile goes from the line for the lower extreme to the bottom of the box.

▶ The bottom of the box is at about 30.75, and the lower extreme is at 27. So the range for the lower quartile is 3.75.

Histograms

A **histogram** is another type of chart that allows us to see how numerical data is distributed. It looks like this:

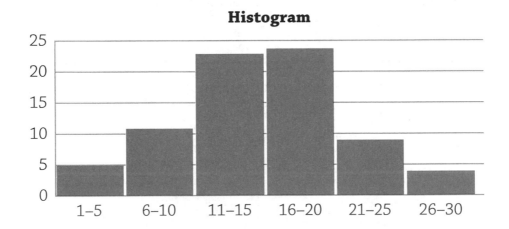

Histogram

At first glance, this may look a lot like a bar graph, but a histogram is actually quite different. A bar graph compares different categories, such as miles versus hours. A histogram, instead, shows the same kind of data as a stem-and-leaf plot or a box-and-whisker plot. It displays only one data set and allows you to see the distribution of that data. Each bar on a histogram shows how many values fall into a particular range. The more values there are in a group, the higher the bar will be.

One great thing about histograms is that they can display a huge amount of data in a single chart. Another great thing is that the person creating the histogram decides what the ranges are. For example, let's do a chart of exam scores with the scores grouped by letter grade. We can use the same set of math test scores we looked at for the box-and-whisker plot: {72, 66, 84, 95, 78, 82, 80, 83, 86, 79, and 90}. First, we group the scores by letter grade: 90–100 is an *A*, 80–89 is a *B*, 70–79 is a *C*, 60–69 is a *D*, and anything below 60 is an *F*. Count how many scores we have for each group. There are 2 *As*, 5 *Bs*, 3 *Cs*, 1 *D*, and no *Fs*. Those numbers will tell us how high to make the bars in our histogram.

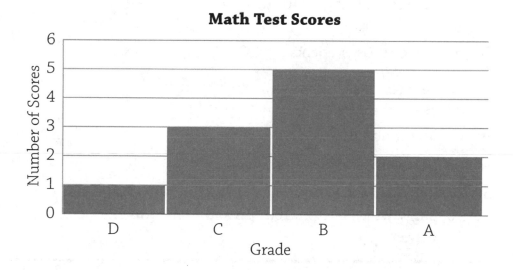

This type of chart allows us to see how the scores are grouped. At a glance, we can see that *B* is the most common grade.

EXAMPLE

▶ The following chart shows the number of absences for a group of 80 high school students over their four years in high school.

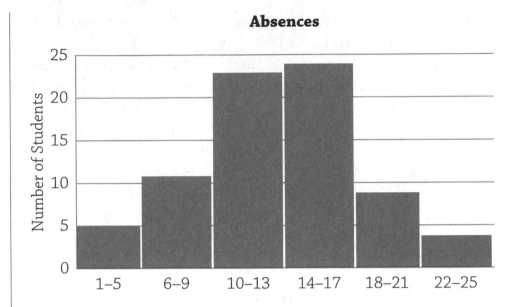

Which range do most students fall in?

    a. 6–9 absences

    b. 6–13 absences

    c. 10–17 absences

    d. 18–25 absences

▶ This histogram shows that the bulk of the data falls into the middle two groupings, which go from 10 to 13 and from 14 to 17. Choice *C* is the best answer.

Tree Diagrams

A **tree diagram** shows the possible combinations of things. It is useful for calculating probabilities, permutations, and combinations, which we will do in the next chapter.

Here is a tree diagram showing the possible outcomes when a coin is flipped twice:

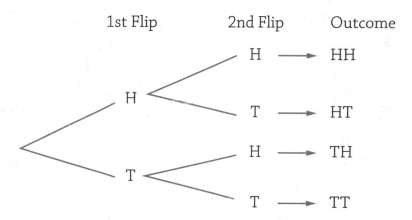

Each time the coin is flipped, there are two possible outcomes: heads or tails. The first branch shows those two possible outcomes labeled *H* and *T*. When the coin is flipped a second time, there are two possible outcomes: heads or tails. For each of the two first outcomes, we draw a branch showing the two possible outcomes of the second flip.

This branching is why we call this a tree diagram. Each time we branch off, the diagram gets bigger and more complicated, so it is important to go slowly and be careful and thorough when drawing one of these diagrams. At the far right of the diagram, you can see a list of all the possible outcomes. Each one shows the path followed along the branches to get to that end point. We can count up those endpoints to see how many possible outcomes there are. In this case, there are four.

EXAMPLE

▶ At a wedding dinner there are three choices of entree: beef, chicken, or fish. There are four choices of side dish: salad, pasta, rice, or mixed vegetables. How many different combinations are possible? Draw a tree diagram.

▶ Let's use letters to indicate the entrees. Then add the group of side dishes to each of them:

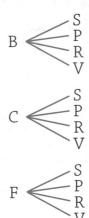

▶ Count the end branches. There are 12 possible combinations.

Venn Diagrams

A **Venn diagram** shows how two or more categories of data overlap. A Venn diagram is usually shown by overlapping circles, with each circle representing a category of data. A Venn diagram with two categories looks like this:

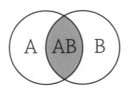

The space where the two categories overlap is for data that falls into *both* categories.

A Venn diagram with three categories looks like this:

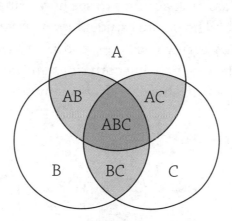

If we try to group people into only two groups, such as people who have a cat and people who have a dog, we have a hard time showing the people who have both a cat *and* a dog. With only two groups, we would have to put those people into both of the groups, and that would mess up the total because those people were counted twice. A Venn diagram allows us to overlap the two groups and see the people who have both pets clearly, while still maintaining the correct total number of people. The diagram below shows this diagram for 30 people who have cats, 25 people who have dogs, and 10 people who have both.

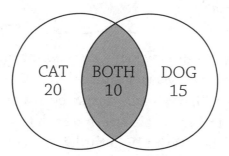

The numbers listed in the diagram show, in the left circle, that 20 people have only a cat and 10 people have a cat and a dog. The entire area of the cat circle adds up to 30. The right circle shows that 15 people have only a dog and 10 people have both a dog and a cat. The entire area of the dog circle adds up to 25.

EXAMPLE

Mr. Eichenberg's class took a survey to see how many students were wearing necklaces and how many students were wearing rings. Twenty-three students were wearing necklaces, and 12 students were wearing rings. Four students were wearing both rings and necklaces. Fill in the Venn diagram below with this data. How many students are wearing only a ring?

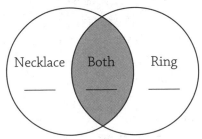

The overlapping area is easy, so let's fill that in first. The question states that 4 students are wearing both, so write a 4 in the shaded overlapping area.

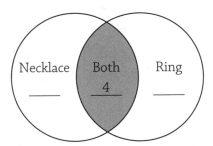

Now we need to figure out how many are wearing only a necklace and how many are wearing only a ring. Since the total number of people wearing necklaces is 23, but 4 are wearing both a necklace and a ring, we should subtract the 4 who are wearing both to get 19 who are wearing only a necklace. Write 19 in the left circle.

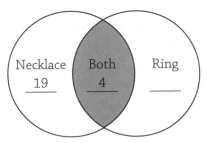

The total number of people wearing rings is 12, but 4 are wearing both a ring and a necklace, so should subtract the 4 who are wearing both to get 8 who are wearing only a ring. Write 8 in the right circle. Eight students are wearing only a ring.

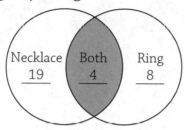

Comparing Data Distributions

Line graphs and bar graphs allow us to see the data for more than one person or thing all in one graph. Other types of graphs can show only one data set. To compare data, you need more than one graph.

If two kids (who totally love math with visual aids) want to compare the types of things they have in their toy boxes, they could each make a circle graph and then compare them.

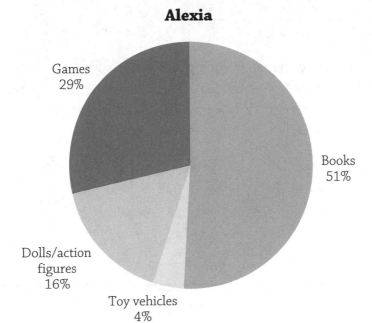

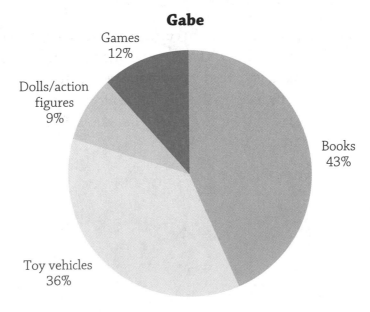

We can see that the same four categories are shown in both graphs: games, books, toy vehicles, and dolls/action figures. Here are some comparisons we might make:

- For both children, books make up the largest percent of the items in their toy boxes.

- For both children, books make up roughly half of the total number of items.

- Alexia has a higher percentage of games than does Gabe.

- Gabe has a higher percentage of toy vehicles than does Alexia.

When we are creating data displays, we can choose which type of graph best suits our purpose. In other words, what type of graph makes it easiest to see what we want viewers of the graph to see? Take a look at these two graphs of student grades on a history test. The first graph is a box-and-whisker plot. The second is a stem-and-leaf plot.

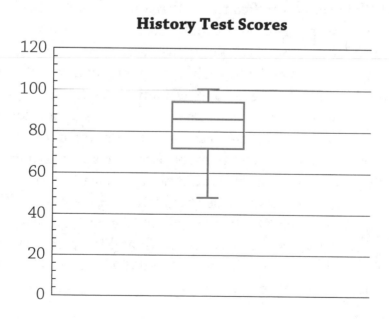

History Test Scores

Stem	Leaf
1	
2	
3	
4	8
5	
6	7, 8, 9
7	2, 2, 3, 9
8	1, 6, 6, 6, 7, 9
9	0, 4, 4, 4, 5, 5, 6
10	0

Which graph makes it easier to see the median score? The box-and-whisker plot makes it easier to see the median because it has a line marked at the median, which is 86. We can find the median on the stem-and-leaf plot by counting how many scores there are and then identifying the middle one, but the box-and-whisker plot makes finding the median easier.

Which graph makes it easier to see how many students scored an 86? The stem-and-leaf plot makes finding out how many students made a particular score because it lists each individual score. We can go to the row for stem 8 and then count how many 6s there are in the leaf column. Three students scored an 86. The box-and-whisker plot cannot be used to answer this question at all because it shows the data distribution but not the individual scores.

Which graph makes it easier to see the highest score on the test? This can be done with either graph. On the box-and-whisker plot, all we need to do is find the top line, which is at 100. On the stem-and-leaf plot, all we need to do is look at the highest stem number, 10, and the highest leaf number, 0, to see that the highest score on the test is a 100.

EXAMPLE

▶ Below are two histograms that show the high temperatures for August and how many days each city experienced those temperatures. Graph *A* shows Houston, Texas. Graph *B* shows New York, New York. Which city had the larger range of temperatures? Which city had the highest average temperature?

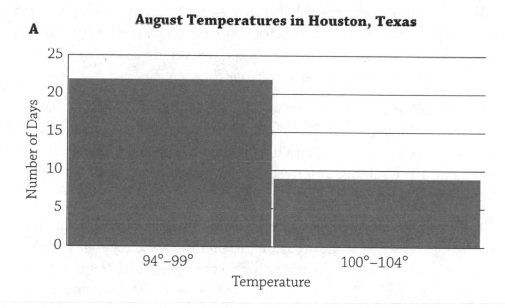

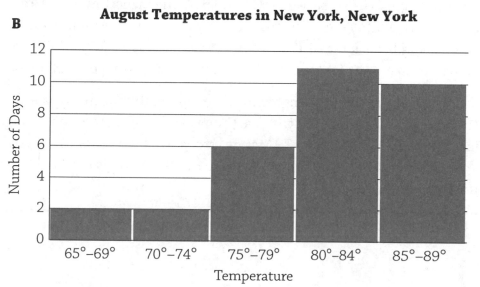

▶ After looking at the two graphs, it seems like New York has a larger range of temperatures. The highest possible temperature is 89°, and the lowest possible temperature is 65°. That's a difference of 24°. In Houston, the lowest possible temperature is 94°, and the highest possible temperature is 104°. That's a difference of 10°.

▶ We have to be careful, though, because each bar of a histogram shows a range, so we do not know the exact temperature for each day. In order to find the exact range for each data set, we would need to know the actual highest and lowest temperatures for each data set.

▶ For example, on the graph for New York, the first bar shows us that two days were between 65° and 69°. That means the lowest *possible* temperature was 65°, but both days could have been 69° and the bar would look the same. Our lowest temperature, then, is somewhere between 65° and 69°.

▶ Similarly, the highest *possible* temperature in New York is 89°, but all we really know is that the actual highest temperature is between 85° and 89°. The range from lowest to highest temperatures, then, could be as low as 16 (85°–69°) or as much as 24 (89°–65°). For Houston, the lowest temperature is between 94° and 99°. The highest temperature is between 100° and 104°. The range from lowest to highest temperatures, then, could be as low as 1 (100°–99°) or as much as 10 (104°–94°). In any case, the range for New York is greater than the range for Houston.

▶ Finding the city with the higher average temperature is much easier, though we still cannot tell the exact number. For Houston, the average is going to be in the 94° to 99° range, probably closer to 99°. For New York, the average is going to be closer to 80°. Houston is definitely hotter!

EXERCISES

EXERCISE 9-1

Use the bar graph to answer questions 1–3.

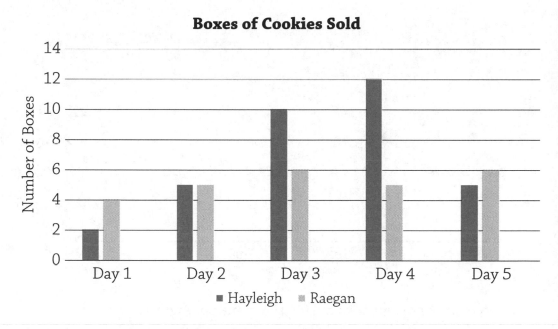

Boxes of Cookies Sold

1. Who sold more boxes of cookies on Day 5?

2. On which day did Hayleigh and Raegan sell the same number of boxes?

3. Who sold the most boxes overall?

EXERCISE 9-2

Use the line graph to answer questions 1–3.

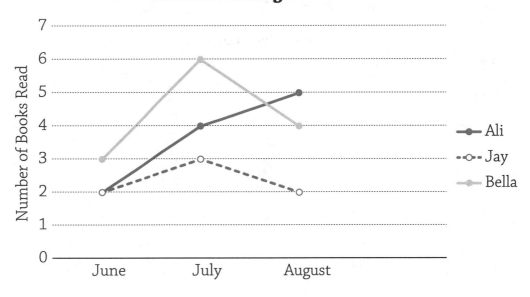

Summer Reading Contest

1. Who read the most books during the month of August?

2. During which month did two people read the same number of books?

3. Which person showed an increase in the number of books read each month?

EXERCISE 9-3

Use the circle graph to answer questions 1–3.

Monthly Expenses

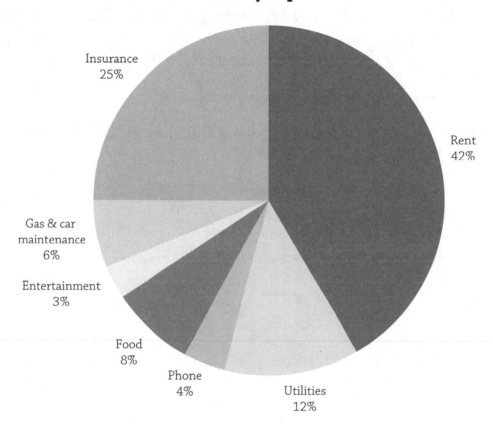

1. Which type of expense makes up the smallest share of the monthly expenses?

2. Which two expenses, added together, are the same as utilities?

3. If the total amount of expenses per month was $2,000, what amount was spent on insurance?

EXERCISE 9-4

Let's put what we've learned about stem-and-leaf plots to use with the following questions.

1. The Bethesda High School varsity basketball team played 14 games this year. Their scores were: 36, 47, 62, 55, 49, 68, 63, 42, 59, 61, 50, 48, 45, and 55. Create a stem-and-leaf plot that shows this data.

Use the stem-and-leaf plot here to answer questions 2–4.

This chart shows the points scored by the Bethesda High School varsity basketball team in each game last year.

Stem	Leaf
1	
2	
3	5, 7
4	0, 6, 7, 7
5	3, 3, 8, 8, 9
6	2, 7

2. What is the range of the scores?

3. What is the median score?

4. In how many games did the team score more than 60 points?

EXERCISE 9-5

Use the box-and-whisker plot to answer questions 1–3.

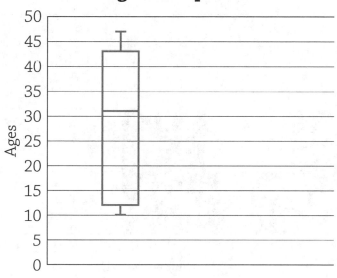

Ages of People on Bus

1. What is the median age for the people on the bus?

2. What is the age of the youngest person on the bus?

3. What is the range of ages for the 3rd quartile?

EXERCISE 9-6

Use the histogram below to answer the following questions.

Exercise Class Sit-Ups

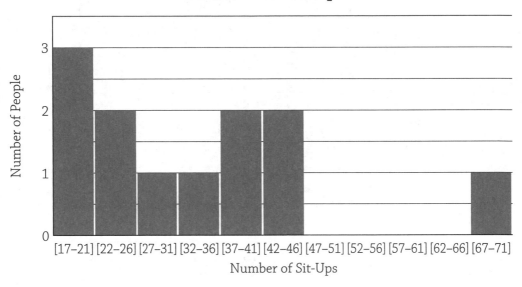

1. How many people did more than 50 sit-ups?

2. How many people did between 27 and 36 sit-ups?

3. What is the total number of people in the class?

EXERCISE 9-7

Use the information below to answer these questions about tree diagrams.

David puts three marbles—one red, one white, and one black—into a bag. He draws one marble, records the color, and then puts it back. He draws a marble a second time, records the color, and puts it back. The tree diagram here shows the possible outcomes. Use the tree diagram to answer questions 1–2.

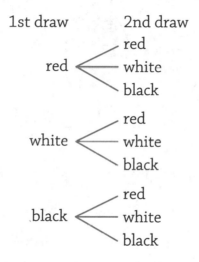

1. How many possible outcomes are there?

2. How many outcomes are there in which the same color marble is drawn both times?

3. Shannon wants to order a pizza. She can choose between thick or thin crust and can have pepperoni, sausage, or plain cheese for toppings. Draw a tree diagram to show her choices.

EXERCISE 9-8

Use the Venn diagram below to answer these questions.

This diagram shows the results of a survey in which people were asked whether they can roller skate, ice skate, or both. Forty-nine people said they can roller skate, 36 people said they can ice skate, and 11 people said they can do both.

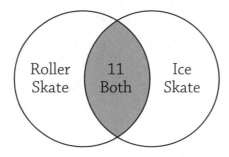

1. How many people can *only* ice skate?

2. What is the total number of people surveyed?

3. Draw a Venn diagram that shows the results of a survey in which 100 people were asked whether they can speak French, Spanish, or German. Thirty people speak German, 38 people speak Spanish, and 32 people speak French. Two people speak all three languages, 6 people speak both French and German, 4 people speak both German and Spanish, and 12 people speak both French and Spanish.

EXERCISE 9-9

Use the two box-and-whisker plots here to answer this group of questions. Both graphs show the same data.

A

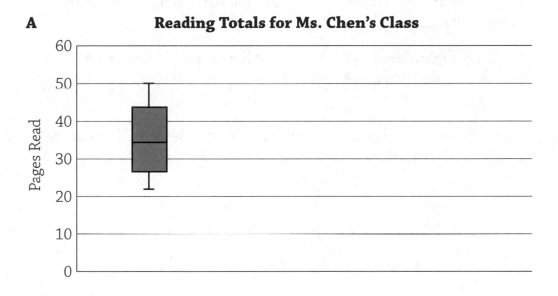

Reading Totals for Ms. Chen's Class

B

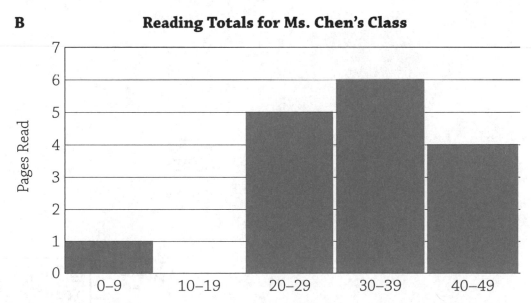

Reading Totals for Ms. Chen's Class

1. Which graph most clearly shows that one or more students read no pages?

2. Which graph shows that no students read between 10 and 19 pages?

3. Which graph makes it easier to find the median number of pages read? What is the median?

4. Which graph makes it easier to find out how many students read between 30 and 39 pages? How many students read between 30 and 39 pages?

5. Which chart would you use to find the total number of students in the class? How many students are there?

10 Statistics

MUST ⚡ KNOW

⚡ Statistics help us sort and analyze data and allow us to make more accurate predictions about the world around us.

⚡ The four most common measures of central tendency are: mean, the average of a data set; median, the middle number of a data set; mode, the most frequently occurring data point; and range, the difference between the largest and smallest values in a data set.

⚡ Probability is a measure of the likelihood that a particular event will occur. It is the total number of possible outcomes divided by the number of times a particular event occurs.

I n the last chapter, we talked a lot about data and ways to represent data. We looked at box-and-whisker plots and histograms that show how data is distributed across a range of values. In this chapter, we will continue our study of data distribution by learning how to find measures of central tendency such as the mean, median, and mode. We will also look at how data are gathered and how to predict future data through probability. These topics in mathematics are called **statistics**. Statistics is the study of collecting, presenting, and analyzing data.

Sampling

A **census** is an official count and survey of characteristics of the entire population. When the United States does a census, it collects data from every person in the entire country. This process is so difficult and time-consuming that it is done only once every ten years.

When scientists or companies want to collect data, they usually start by identifying the population for the things they want to study. In statistics, a **population** refers to the whole group that we are interested in studying. Many populations are much too large to survey or study, so we take a **sample**, which is a collection of data from just *some* members of the population. Sampling always runs the risk of missing some data since not every person in the population is included, but it is the most practical way to gather data. The more people chosen for the sample group, the better the sample will be.

There are four basic methods of sampling:

- **Random**   A random sample chooses people blindly, without knowing anything at all about them. If we wanted a random sample of students at a school, we could write each student's name on an identical piece of paper, put all the papers in a bag, mix them up, and draw out however many names we wanted to use.

■ **Systematic**   A systematic sample has a method for choosing participants, such as choosing every fourth person. This is similar to random sampling because nothing is known about the participants when they are chosen. If we wanted a systematic sample of students at a school, we could get a list of all the students' names and circle every fourth name. Those circled will be the participants.

■ **Stratified**   A stratified sample aims to make sure that the population is represented fairly in the sample group. To do this, the population is divided into categories, perhaps by age or gender or race. Whatever percentage of the population falls into those categories, the sample must have the same percentage for the same categories. If we wanted a stratified sample of students at a high school, we might divide the students by grade level. If we find 18% are seniors, 22% are juniors, 25% are sophomores, and 35% are freshmen, then we apply those same percentages to the sample group we choose. We might then use a random sample of each grade level to choose the proper number of names for each grade level.

■ **Cluster**   A clustered sample divides the population into many groups and then randomly chooses entire groups. If we wanted a clustered sample of students at a school, we could divide the students by their homeroom class and then randomly choose five homeroom classes to participate. This type of sample runs a higher risk of being inaccurate, since portions of the population may not be represented at all. It should be used only when the groups are very similar in their characteristics.

EXAMPLE

▶ Which of the following is an example of a stratified sample taken from the employees at a corporation that has 500 employees?

    a. A computer program randomly chooses 50 names from a database of the company's employees.

> b. The human resources manager randomly chooses five people from each of the ten departments at the company.
> c. The company CEO sends the survey by e-mail to every employee at the company.
> d. The human resources manager randomly chooses 20 of the 200 female employees and 30 of the 300 male employees.

▶ A stratified sample is done based on some characteristic of the population, such as age or gender, and uses the same percentages of that characteristic for the sample group as there are in the total population.

▶ Choice d is an example of that. The company has $\frac{200}{500}$ or 40%, female employees and $\frac{300}{500}$ or 60%, female employees. Of the 50 people chosen for the survey, the manager has chosen $\frac{20}{50}$, or 40%, female and $\frac{30}{50}$, or 60%, male employees to survey.

The goal of any method of sampling is to get a **representative sample** of the entire population. This means that the data on the sample group is the same as it would be in a study of the entire population. To get a representative sample, it is important to avoid **bias**. Bias is the tendency of a statistic to overestimate or underestimate results. It can be due to measurement errors or the way the sample group is chosen. For example, in a survey of students at a school, if only female students are chosen, the results may not accurately reflect the entire population of the school. There are many types of bias, but here are a few of the most common:

- **Selection bias**   This happens when the sample is not representative of the population due to how the sample group was chosen. An example of this would be a survey done to measure student satisfaction with gym facilities that only surveyed the student council members.

- **Self-selection bias** This is similar to selection bias, but instead of the surveyor selecting participants, here the participants choose themselves. An example of this would be product reviews. People choose to leave reviews of products, and those who do are more likely to be very happy or very disappointed. This means the product rating may not be representative of everyone who uses the product.

- **Recall bias** Recall bias occurs because human memory changes over time. If a survey is done about events that occurred in the past, the people surveyed may remember only the good parts of the event or just the horrible thing that went wrong.

- **Observer bias** This type of bias occurs when the surveyor (the person giving the survey) projects his or her opinion onto the subjects, whether consciously or unconsciously. If the surveyor has a bias, it is likely to come through in the way questions are phrased or in the way survey results are interpreted by the surveyor. An example of this would be a teacher who has students evaluate his performance while he is in the room or lets the students know that he will be reading the evaluations.

EXAMPLE

▶ You are designing a survey to determine whether students at your school want more school-sponsored clubs. Which of the following sample groups is the most representative of the population?

       a. students who are not currently involved in a club
       b. students who are currently involved in a club
       c. twenty students in each grade level, chosen at random
       d. first-year students

▶ Since the survey is meant to determine whether students at the school want more clubs, we need to survey all the students at the school. Since we are only surveying a sample of those students, it needs to be representative of the entire group of students.

> Choices *a*, *b*, and *d* are subsets of the population and may not be an accurate representation of the entire population. Choice *c*, since it is a systematic random sample, is most likely to be a representative sample of the entire student population.

Measures of Central Tendency

Now that we have discussed the issues involved in gathering accurate, unbiased data, we can move on to discussing what to do with data after it has been gathered. Since researchers are usually concerned with what the "average" person thinks or does, the middle of the data distribution is important. There are several ways to measure **central tendency**, which is the center or most typical value. We tend to think of this as the average, but there are other ways to measure the center as well. We will look at averages, as well as the median, mode, and range of data sets.

Arithmetic Mean

The **arithmetic mean** is what we commonly refer to as the average of a set of data. We find the mean by adding up all the values and then dividing that sum by the number of values we added. To find the average of the data set {5, 8, 11, 4}, we add up the values ($5 + 8 + 11 + 4 = 28$) and then divide that by 4 because there were 4 values in the data set: $28 \div 4 = 7$. The arithmetic mean of the data set, then, is 7.

EXAMPLE

> Isaac wants to know his current grade in social studies. His grade is the average of six unit-test scores, and he has taken five of them so far. His scores are: 75, 80, 90, 83, and 87. What is his current average?

▶ To answer this question, we need to find the arithmetic mean (average) of the five scores he has so far. Add up the scores: $75 + 80 + 90 + 83 + 87 = 415$.

▶ Since there are five scores, we divide the total by 5: $415 \div 5 = 83$. Isaac's current grade is an 83.

Median

The **median** is the middle value of a data set. We worked with median a bit in the last chapter when we discussed box-and-whisker plots. The most important thing to remember about finding a median is that the data set values must be put in order before we can identify the middle value. Let's find the median of the data set from the last example (Isaac's grades). His grades were 75, 80, 90, 83, and 87. First, we need to put the values in order: {70, 80, 83, 87, 90}. Now we can find the middle value. Since there are five values, the third one will be the median: 83. In this case, the mean and the median are exactly the same!

BTW

When I find a median, I like to write the list of numbers, then cross off numbers on either end, moving toward the center, until I get to the middle:

1, 7, 11, 15, 19, 22, 25, 29, 30, 34, 37

EXAMPLE

▶ Olivia wants to know the heights of the players in the starting lineup for the Los Angeles Lakers game she is going to watch. She gets a list of the players' heights from the team website and records them in inches:

LeBron James	80 inches
Anthony Davis	82 inches
JaVale McGee	84 inches
Danny Green	78 inches
K. Caldwell-Pope	77 inches

What is the median height? What is the average height?

▶ To find the median, we need to put the values in order: {77, 78, 80, 82, 84}.

▶ Now, we identify the middle value. Since there are five values, the middle one is the third value. The median height is 80 inches.

▶ To find the average, we need to add up the values: 80 + 82 + 84 + 78 + 77 = 401.

▶ Now we divide by the number of values (5): 401 ÷ 5 = 80.2.

▶ The average height is 80.2 inches.

What happens if there are an even number of values? Then there isn't a "middle" number. In that case, we take the average of the two numbers in the middle.

EXAMPLE

▶ Isaac takes his sixth unit test and scores a 92. His other scores are: 75, 80, 90, 83, and 87. What is the median score? What is his new grade average?

▶ To find the median, we put the values in order: {70, 80, 83, 87, 90, 92}.

▶ Now we identify the middle value. In this case, there are two middle values: 83 and 87. We need to average those values to find the median: 83 + 87 = 170 and 170 ÷ 2 = 85. The median is 85.

▶ To find the average of the six scores, we add up the scores: 75 + 80 + 90 + 83 + 87 + 92 = 507. Since there are six scores, we divide the total by 6: 507 ÷ 6 = 84.5.

▶ Isaac's grade is an 84.5, which may be rounded to 85.

Mode

The **mode** is the most frequently occurring value in a data set. If every value in a data set is different, we say that there is no mode. For the data set of Isaac's grades, there is no mode. For the data set {3, 6, 8, 8, 11, 14}, the value 8 appears twice while all other values appear only once. That makes 8 the mode of the data set.

Mr. Smith surveys his 7th grade class to find out each student's age. He records the results in a bar graph. What is the mode of the data set?

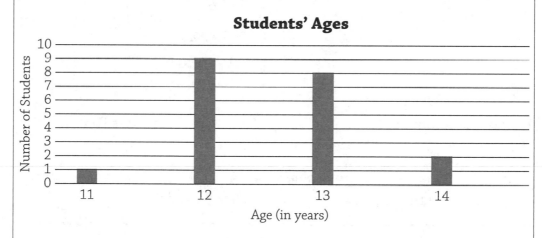

To find the mode of the data set, we need to find the most common age. The graph shows ages on the horizontal axis and the number of students along the vertical axis.

The tallest bar will be the most common age. The tallest bar is for 12-year-old students (there are nine of them). The mode is 12.

What if more than one value occurs the most? In this case, there is more than one mode.

Dr. Meinhart studies the diastolic blood pressure readings of her patients who are currently in the hospital. She records the results in a stem-and-leaf plot. What is the mode of the data set?

Stem	Leaf
6	5
7	7, 9, 9
8	0, 2, 5, 5, 5, 6, 8, 9
9	3, 3, 3, 4, 6, 8
10	0, 2, 4
11	5

▶ Look for numbers that repeat: 79 appears twice, 85 appears three times, and 93 appears three times.

▶ Three times is the most any value appears, but we have two values that appear three times, so there are two modes: 85 and 93.

Range

Range is the spread of the data set. We find range by subtracting the smallest value from the largest value. In the last chapter, we found the range on some of our data displays, so you should already be familiar with this measure of central tendency. We do need to be sure that we have the absolute smallest and largest values, so it is helpful if the data is in order.

Let's look at the range of Isaac's test scores. His six test scores, in order, are: {70, 80, 83, 87, 90, 92}. The largest number is 92, and the smallest is 70, so the range is: $92 - 70 = 22$.

> Olivia continues to look at Lakers statistics by making a chart of the number of points LeBron James scored in each preseason game. What is the range of this data set?

Opposing Team	Points Scored
Golden State Warriors	18
Brooklyn Nets	6
Brooklyn Nets	20
Golden State Warriors	15

> To find the range, we need to find the largest and smallest values. The largest value is 20, and the smallest is 6, so the range is: $20 - 6 = 14$.

Mean Absolute Deviation

Range gives us a number that shows the spread of the data from the smallest to the largest value. Arithmetic mean gives us an average of all the values. Deviation means how far a data point is away from the average. **Mean absolute deviation** is the average (mean) of the distance away from the average (deviation) for each data point.

It's a bit of a pain to find the mean absolute deviation, but it is a very helpful thing to know when we are evaluating data distribution. The measures of central tendency don't always give us a good mental picture of the data distribution. The mean and median of {49, 50, 51} are both 50, but so are the mean and median of {1, 50, 100}. Those are very different distributions!

Knowing the range of those values would help us see that. The range of {49, 50, 51} is 3, while the range of {1, 50, 100} is 99. Even knowing both the mean and the range doesn't always give us the whole picture though. Mean absolute deviation can give us a more thorough idea of how the data is spread.

To find the mean absolute deviation of {49, 50, 51}, we first find the mean: 50. Now we calculate how far away from the mean each of the values lies. We do this with absolute value, so there will be no negative distances.

Value	Distance from Mean
49	$50 - 49 = 1$
50	$50 - 50 = 0$
51	$|50 - 51| = 1$

Now we average the distances: $\dfrac{1+0+1}{3} = \dfrac{2}{3}$. This tells us that the average distance from the mean is only $\dfrac{2}{3}$. The data are very close together.

For the other data set, we find a very different value for the mean absolute deviation.

Value	Distance from Mean
1	$50 - 1 = 49$
50	$50 - 50 = 0$
100	$|50 - 100| = 50$

Now we average the distances: $\dfrac{49+0+50}{3} = \dfrac{99}{3} = 33$. With a mean absolute deviation of 33, we know that this data set is spread widely.

EXAMPLE

▶ Isaac's test scores on his six unit-tests are 75, 80, 90, 83, 87, and 92. Find the mean absolute deviation.

▶ First, we find the mean. Add up the scores: $75 + 80 + 90 + 83 + 87 + 92 = 507$. Since there are six scores, we divide the total by 6: $507 \div 6 = 84.5$.

▶ Now we need to find how far each value lies from that mean.

Value	Distance from Mean
75	$84.5 - 75 = 9.5$
80	$84.5 - 80 = 4.5$
90	$|84.5 - 90| = 5.5$
83	$84.5 - 83 = 1.5$
87	$|84.5 - 87| = 2.5$
92	$|84.5 - 92| = 7.5$

▶ Now we average the distances:

$$\frac{9.5 + 4.5 + 5.5 + 1.5 + 2.5 + 7.5 = 31}{5} = 6.2$$

▶ With a mean absolute deviation of 6.2 points, we know that Isaac's grades are not very far apart. We can also see that his score of 75 is a bit of an **outlier** (a point far from the rest of the data) since it has the greatest deviation from the mean, nearly 10 points.

IRL Sports coaches can use mean absolute deviation to select players. Look at the data for these two basketball players:

	Vivian	Evelyn
Game 1	84	62
Game 2	68	55
Game 3	32	58
Game 4	56	65

The average number of points scored by both players is 60, so we might think they are alike. When we look at the mean average deviation (MAD), however, we see that Vivian's MAD is 13.5 while Evelyn's is only 3.75. Evelyn is a much more consistent player, so if consistency is important to a coach, the coach would pick Evelyn over Vivian.

Probability

Now that we have discussed ways to analyze data, let's talk about how to *predict* data. In mathematics, we call this **probability**. Probability is the likelihood that something will occur. We use probability every day, even if we are unaware of it. We look at the weather forecast to see that there is a 20% chance of rain today. Sports teams and fans both look at past statistics to predict how a certain player or team will do in a future game. Companies use probability to project future sales and earnings. Insurance companies use probability to determine levels of risk.

Probability can be expressed as a fraction or decimal. A 100% probability or a probability of 1 means that an event will absolutely occur. A probability of 0 means there is absolutely no chance of the event occurring. To calculate the probability of an event occurring, we divide the number of times the event we are looking at could occur by the total number of possible outcomes.

That sounds complicated, but it really is not. If we want to find the probability of getting "heads" when we toss a two-sided coin (heads/tails), we take the number of possibilities for heads, which is 1 since a coin has only one "heads" side, and divide that by the total number of possible outcomes, which is 2 because the coin has two sides. The probability of getting heads when we toss a coin, then, is $\frac{1}{2}$, or 50%. The same is true for the probability of getting "tails." One tails out of two possible outcomes is $\frac{1}{2}$, or 50%.

Coin tossing is a popular topic for probability questions. Rolling dice and drawing a card from a deck of cards are also popular, so let's make sure we know the basics of how those work. A die is a single, numbered cube, typically numbered from 1 to 6. A deck of cards has 52 cards, with four different types of cards, called suits: hearts, spades, diamonds, and clubs. Each suit has 13 cards: the numbers from 1 to 10, plus a jack, a queen, and a king. Anything can be a topic for a probability question, but these three are common.

EXAMPLE

▶ What is the probability of rolling a 4 on a six-sided die, numbered from 1 to 6?

▶ The die has six sides. Each side has one of the numbers from 1 to 6. How many numbers on the die are a 4? One. How many numbers are there total? Six. The probability of rolling a 4, then, is $\frac{1}{6}$.

What if there is more than one option for the outcome we are wondering about? If we have a bag with four marbles and two are red and two are blue, then we would have more than one option for drawing out a blue marble. We do this the same way, however. There are two blue marbles out of four total marbles, so the probability of drawing a blue marble is $\frac{2}{4}$. We can reduce that to $\frac{1}{2}$.

▶ What is the probability of rolling an even number on a six-sided die, numbered from 1 to 6?

▶ There are six total possible outcomes since there are six sides. How many numbers on the die are even? Three: 2, 4, and 6.

▶ The probability of rolling an even number is $\frac{3}{6}$. We can reduce that to $\frac{1}{2}$.

What if we want to find the probability that something will *not* occur? What if we are playing a game and rolling a four would be bad? We can calculate the probability of something not happening in the same way that we do for the probability that something will happen. Count the number of times the event does *not* occur and divide that by the total number of possible outcomes. If we want the probability of *not* getting "heads" when we toss a coin, we can say that there is one outcome that is not heads (tails) and that there are two total possible outcomes. The probability of *not* getting heads, then, is $\frac{1}{2}$.

▶ What is the probability of *not* rolling a 4 on a six-sided die, numbered from 1 to 6?

▶ The die has six sides. Each side has one of the numbers from 1 to 6. How many numbers on the die are *not* a 4? Five. How many numbers are there total? Six. The probability of not rolling a 4, then, is $\frac{5}{6}$.

There is another way to find the probability that an event will not occur. Notice that the probability of rolling a 4 is $\frac{1}{6}$ and the probability of not rolling a 4 is $\frac{5}{6}$. There are no other ways rolling a die can go. We either get a 4 or we don't, right? Since those are all the possible outcomes, the probability of those outcomes adds up to 1: $\frac{1}{6} + \frac{5}{6} = 1$.

Similarly, we found that there is a probability of $\frac{1}{2}$ for tossing "heads" on a coin and a probability of $\frac{1}{2}$ for tossing "tails" (*not* heads). Those are the only two options when we toss a coin. If we add those probabilities up, we get: $\frac{1}{2} + \frac{1}{2} = 1$.

Since probability is a fraction (from 0 to 1), and 1 is the absolute probability that the event will occur, the probability of an event occurring plus the probability of the event not occurring *always* adds up to 1. Knowing this allows us to find the probability of an event not occurring by subtracting the probability that it will occur from 1. If the probability of rain is 20%, for example, then the probability that it will not rain is 80%.

EXAMPLE

▶ What is the probability of *not* drawing a king from a deck of 52 cards?

▶ For this type of question, we might decide it is easier to find the probability of drawing a king and subtracting that from 1. How many kings are there? Four, one of each suit. There are 52 total cards, so the probability of drawing a king is $\frac{4}{52}$.

▶ The probability of not drawing a king, then, is $1 - \frac{4}{52}$. We can write 1 as $\frac{52}{52}$, so we have: $\frac{52}{52} - \frac{4}{52} = \frac{48}{52}$. That can be reduced to $\frac{12}{13}$.

We can find the probability of multiple events occurring by finding the individual probability for each event and then multiplying them together. If we flip a coin twice and want to know the probability of getting heads both times, we find the probability of getting heads on the first flip ($\frac{1}{2}$) and the probability of getting heads on the second flip (also $\frac{1}{2}$), because the second flip really doesn't have anything to do with the first flip: heads is still one of two possible outcomes. Then we multiply those together: $\frac{1}{2} \times \frac{1}{2} = \frac{1}{4}$.

▶ What is the probability of drawing a king, putting the card back in the deck, and then drawing a king again from a deck of 52 cards?

▶ How many kings are there? Four (one of each suit). There are 52 total cards, so the probability of drawing a king on the first draw is $\frac{4}{52}$, or $\frac{1}{13}$.

▶ Since we put the card back, the probability of drawing a king on the second draw is also $\frac{1}{13}$. We multiply those together to find the probability that both will occur: $\frac{1}{13} \times \frac{1}{13} = \frac{1}{169}$. Not great odds!

What would happen if we didn't put the first king back in the deck? That really changes things. The probability of drawing a king on the first draw is $\frac{4}{52}$, or $\frac{1}{13}$. If we don't put the card back, then there are only 51 cards and only 3 kings left. The probability of drawing a king on the second draw is now $\frac{3}{51}$, or $\frac{1}{17}$. We multiply the two probabilities together to find the probability that both will occur: $\frac{1}{13} \times \frac{1}{17} = \frac{1}{221}$.

Using Graphs for Probability

We can use several of the data presentation graphs we learned about in the last chapter to calculate probabilities. Let's look at a circle graph that shows the results of a survey of 55 people who were asked what kind of pet they have:

Pet Survey

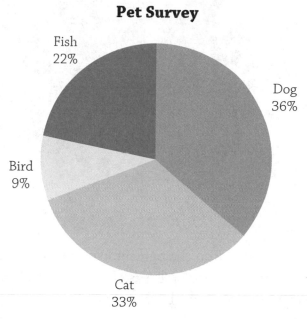

Since a circle graph shows the percentage out of 100 for each category, this can be directly translated into probability. If 9% of the people have a bird, then the probability that a person in the survey has a bird is also 9%.

> The following graph shows the percent of the population in a major city that uses various types of public transportation. What is the probability that a person chosen at random in that city uses either a subway or train?

Public Transportation

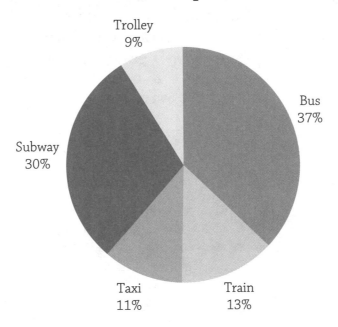

▶ All we need to do here is find the percentage for train use and for subway use and add them up: 13% + 30% = 43%. The probability of a person using either a train or subway is 43%.

Bar graphs and line graphs may also be used to find probability. As long as you can calculate the total number of outcomes and the number of outcomes for the result you are looking at, you can use a bar or line graph to determine probability. Let's look at a bar graph:

Favorite Beverage

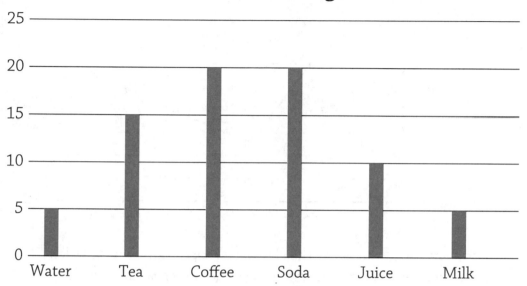

This graph shows the results of a survey of people's favorite beverages. If we want to find the probability that a random person in this group chose milk as a favorite beverage, we need to find the number of people who chose milk and the total number of people surveyed. Find the bar for milk: 5 people chose milk. To find the total number of people surveyed, we need to add up the total for each bar shown: $5 + 15 + 20 + 20 + 10 + 5 = 75$.

The probability that a person in this group chose milk is $\frac{5}{75}$, which can be reduced to $\frac{1}{15}$. If this survey was done on a representative sample of the population, we can say that the probability of milk being someone's favorite beverage is approximately 7% for the whole population.

EXAMPLE

▶ According to the data in the line graph on the next page, what is the probability that the class sold more than 10 raffle tickets in one day?

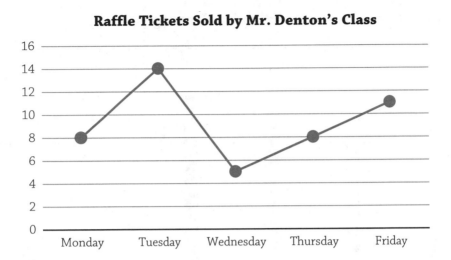

Raffle Tickets Sold by Mr. Denton's Class

▶ To find the probability that the class sold more than 10 raffle tickets in one day, we need to see how many days the class sold more than 10 tickets: 2 days (Tuesday and Friday).

▶ We divide that by the total number of days: 5. The probability that the class sold more than 10 raffle tickets in one day is $\frac{2}{5}$, or 40%.

Any stem-and-leaf plot can be used to find probability since it shows every data point. The next plot shows the number of grammatical errors Ms. Castanon found in each of her student's final exam essays:

Stem	Leaf
1	3, 4, 6, 7, 7
2	1, 5, 7, 7
3	0, 2, 2, 3
4	0, 0, 1

If we want to find the probability that a student made fewer than 15 errors, we can use the plot to see how many values are under 15, which is

2 (13 and 14), as well as how many total values there are: 16. The probability that a student made fewer than 15 errors is $\frac{2}{16}$, which can be reduced to $\frac{1}{8}$.

EXAMPLE

▶ The chart here shows the ages of the participants in a ballroom dancing class that Ivan is taking. Ivan is 27 years old. If dance partners are chosen at random, what is the probability that Ivan will have a partner who is younger than he is?

Stem	Leaf
1	2, 5, 6
2	1, 5, 7, 7, 8
3	2, 2, 4, 6, 8, 9, 9, 9
4	4, 8, 9
5	3, 6, 6
6	1, 7
7	0
8	
9	

▶ This question is a little tricky since it identifies one member of the group and asks about his partner. There are 25 people total, but only 24 of them can be Ivan's *partner*.

▶ The number of people younger than 27 is five, so the probability that Ivan will have a younger partner is $\frac{5}{24}$.

▶ If Ivan wants to know the percent chance, we can convert the fraction to a percent: $\frac{5}{24} = 0.208\overline{3} \approx 21\%$.

A histogram *might* be used to find probability, depending on what question is asked. Since a histogram shows the distribution of data in ranges, though, it can only be used to find some probabilities. Look at the next histogram of the August temperatures in Houston, Texas:

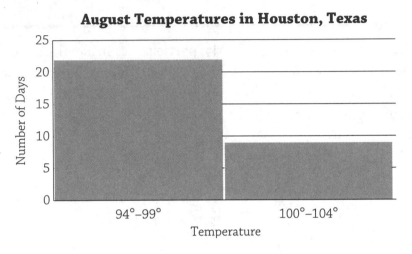

August Temperatures in Houston, Texas

If we want to find the probability that a day in August chosen at random had a temperature below 90°, we can do that. None of the 31 days were below 94°, so the probability is 0. If we want to find the probability that a day was over 100°, we cannot do that because the chart does not divide the temperatures that way. We know that 8 of the 31 days were in the range of 100° to 104°, but we have no idea how many of those 8 days were above 100°.

Remember the tree diagrams we did in the last chapter? Those are very useful for visualizing and calculating probability. If we want to know the probability for getting heads all three times when we flip a coin three times, we can draw a tree diagram of all the possible outcomes (with *H* for "heads" and *T* for "tails").

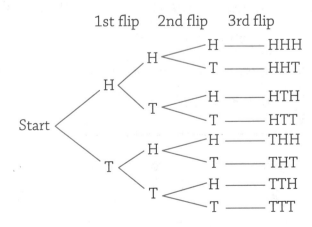

We can count the ends to see that there are eight total possible outcomes. We can follow the pathways to see that only one outcome is heads all three times, so the probability is $\frac{1}{8}$.

EXAMPLE

▶ Mr. Espinol has a bag containing two red marbles and two white marbles. He asks a student to draw a marble, record the color, and then put it back in the bag. The student then draws a second marble and records the color. The tree diagram below shows the possible outcomes. What is the probability that the two marbles drawn are different colors?

1st draw	2nd draw	Outcomes
R	R	RR
	R	RR
	W	RW
	W	RW
R	R	RR
	R	RR
	W	RW
	W	RW
W	R	WR
	R	WR
	W	WW
	W	WW
W	R	WR
	R	WR
	W	WW
	W	WW

▶ We can count the ends to see that there are 16 total possible outcomes. We can follow the pathways to see that there are 8 outcomes in which the marbles are different colors, so the probability is: $\frac{8}{16} = \frac{1}{2}$.

Probability Models

The best type of graph to use to determine probability is one that is specially designed to show probability. We call these **probability models**. When people do surveys or experiments on sample groups, the results can be calculated to show probability. Then those results can be used to predict future events.

Let's learn how to construct a probability table. In a survey of 200 people about their preferred mode of communication, 120 said they prefer texting, 60 said they prefer a phone call, and 20 said they prefer e-mail. We can calculate the probability of each response by using a probability model. We start with a basic table with columns for the modes of communication and the number of people who preferred each one. Then we need a column for the probability. I always like to include a row at the bottom for the totals. This isn't technically necessary, but I find that it helps me keep track of that information and check my calculations.

BTW

I always like to include a row at the bottom for the totals. This isn't technically necessary, but I find that it helps me keep track of that information and check my calculations

Mode of Communication	Number of People	Probability
texting	120	
phone	60	
e-mail	20	
total	**200**	

The probabilities for the three options for the modes of communication have to add up to 1, or 100%. We fill in a 1 for the total probability, and, once we have found all the individual probabilities, we make sure they add up to the total of 1:

Mode of Communication	Number of People	Probability
texting	120	
phone	60	
e-mail	20	
total	**200**	**1**

To find each of the individual probabilities, we take the number of people who chose that mode and divide it by the total number of people (200). For texting, we calculate: $\frac{120}{200} = \frac{12}{20} = \frac{6}{10} = 0.6$. Probability models list probabilities as decimal values, so fill in 0.6 for texting. The probability for phone is: $\frac{60}{200} = \frac{6}{20} = \frac{3}{10} = 0.3$. For e-mail it is: $\frac{20}{200} = \frac{2}{20} = \frac{1}{10} = 0.1$.

Our table should look like this now:

Mode of Communication	Number of People	Probability
texting	120	0.6
phone	60	0.3
e-mail	20	0.1
total	**200**	**1**

Double-check to make sure the probabilities add up to 1. They do. These results can now be used to predict the probability for the larger population (as long as our sample group was a representative sample). We can say that 60% of the population prefers texting, 30% prefers phone calls, and 10% prefers e-mail.

EXAMPLE

▶ Dr. Bumguardner is a geologist who spent several months traveling all over the world looking at rocks and recording the types of rocks she encountered. She classified each rock as sedimentary, igneous, or metamorphic. She found a total of 1,500 sedimentary rocks,

400 igneous rocks, and 100 metamorphic rocks. Calculate the probability for each type of rock in the following table.

Rock Type	Number of Samples	Probability
sedimentary	1,500	
igneous	400	
metamorphic	100	
total	**2,000**	**1**

▶ Divide 1,500 sedimentary rocks by 2,000 total rocks to get 0.75. Divide 400 igneous rocks by 2,000 total rocks to get 0.2. Divide 100 metamorphic rocks by 2,000 total rocks to get 0.1. Dr. Bumguardner's completed table should look like this.

Rock Type	Number of Samples	Probability
sedimentary	1,500	0.75
igneous	400	0.2
metamorphic	100	0.05
total	**2,000**	**1**

Like in the previous example, we can use data tables that show the actual results of an experiment to make predictions about the probability of future events.

EXAMPLE

▶ If Dr. Bumguardner collects 1,000 more rocks, about how many would she expect to be igneous rocks?

▶ Since the probability of a rock being an igneous rock was 0.2 in her previous set of 2,000 samples, she can expect a 0.2 probability of a rock being an igneous rock in her new set of 1,000 samples. A 0.2 probability is the same as 20%, so 20% of 1,000 is 200 igneous rocks.

▶ Of course, this is just a prediction: It doesn't mean she will definitely find exactly 200 igneous rocks out of 1,000. She might find 183 or 218 igneous rocks. The probability just gives her a rough idea of what to expect.

Permutations and Combinations

We need to discuss one more thing before we leave our study of statistics. Often, we want to know how many options we have for something. How many different outfits can we make from our wardrobe? How many different pizzas can we make from a certain number of topping options? How many different ways are there to arrange the seating chart for a wedding? These calculations are called **combinations**, the number of groups that can be formed from a set of options, and **permutations**, the number of ways to arrange items within a group. I remember which is which by thinking of combinations as simply *combining* a bunch of things—the order I put them in does not matter. Permutations are arrangements, so for them, order is everything.

Permutations are actually easier (though larger), so let's do those first. If we want to arrange a group of four candy bars, we can take the number of choices we have for each spot and then multiply those together. We have four choices for spot 1, then three choices for spot 2 (because we have used one of the candy bars, and now only 3 are left from which to choose), then two choices for spot 3, then one choice for spot 1 because just one candy bar is left at this point. Multiply the number of choices: $4 \times 3 \times 2 \times 1 = 24$, so there are 24 ways to arrange four candy bars.

Doing this type of decreasing multiplication is called a **factorial**. A factorial is the product of a given number times each number less than the number down to 1. A factorial is written like this: 4! That exclamation point lets us know that we should take the 4 and multiply it by each number less than 4 down to 1, in other words: $4! = 4 \times 3 \times 2 \times 1$. This is true for any counting number: $6! = 6 \times 5 \times 4 \times 3 \times 2 \times 1$ and $2! = 2 \times 1$.

Combinations are just groups of things, so they are a bit different. If we are just going to eat two of the four candy bars, we may want to know how many different options we have. In other words, how many pairs can we make from a group of four. It does not matter which order we choose them in: {a chocolate bar and a peanut bar} is the same pair as {a peanut bar and a chocolate bar}. If the order did matter, it would be a permutation.

Since it does not matter, this is a combination, and the number of possibilities will be fewer. A permutation would count, say, *AB* and *BA* both, but a combination only counts it once because it's the same pair either way. We find permutations (the number of arrangements) by multiplying together all the choices. In order to make the combination number smaller, we find the permutation and then *divide* it by the number of ways to arrange the items chosen. This sounds harder than it is. If we choose two candy bars from a group of four, we have four choices for the first spot and three choices for the second spot:

$$\frac{4}{1\text{st spot}} \times \frac{3}{2\text{nd spot}}$$

Then we will divide that by the number of ways to arrange the two bars chosen: two choices for the first spot and one choice for the second spot:

$$\frac{(4 \times 3)}{(2 \times 1)}$$

Now, we can reduce before we multiply, which makes the calculations easier.

$$\frac{\overset{2}{\cancel{4}}}{\underset{1}{\cancel{2}}} \times \frac{3}{1} = 6$$

We can choose six different pairs of candy bars from a group of four candy bars. We can also find this from a tree diagram, but those are only practical to use for small numbers of choices. They can get really big really quickly!

We can use mathematical formulas for permutations and combinations. The number of permutations (*P*) of *n* objects taken *r* at a time is determined by the formula:

$$P(n,r) = \frac{n!}{(n-r)!}$$

I find it much easier, though, to just identify the number of choices for each spot and multiply them together.

▶ If Ms. Hanks chooses three framed pictures from a box containing 10 different framed pictures, how many different ways can she arrange them on her shelf?

▶ Identify how many spots we have: 3.

▶ Identify how many choices we have for each spot: 10, then 9, then 8.

▶ Does the order matter? Yes. She is *arranging* them on her shelf. If order matters, we just multiply: $10 \times 9 \times 8 = 720$.

▶ There are 720 ways to arrange 3 pictures chosen from a group of 10. It may take Ms. Hanks a very long time to decide!

The mathematical formula for the number of combinations (*C*) of *n* objects taken *r* at a time is:

$$C(n,r) = \frac{n!}{(n-r)!r!}$$

Again, I find it easier to just identify the number of choices for each spot and then divide by the number of ways to arrange those spots.

▶ If Ms. Hanks chooses three framed pictures from a box containing 10 different framed pictures, how many different groups of pictures could she choose?

▶ Identify how many spots we have: 3.

▶ Identify how many choices we have for each spot: 10, then 9, then 8.

▶ Does the order matter? No. We are looking for *groups*, not arrangements. If order doesn't matter, we need to divide by something.

▶ What do we divide by? The number of ways to arrange the 3 we chose: 3 choices, then 2, then 1.

$$\frac{10 \times 9 \times 8}{3 \times 2 \times 1}$$

▶ We can reduce before we calculate to make it easier.

$$\frac{10 \times \overset{3}{\cancel{9}} \times \overset{4}{\cancel{8}}}{\cancel{3} \times \cancel{2} \times 1} = 120$$

▶ There are 120 ways to group 3 pictures taken from a group of 10. That's a lot less than 720 arrangements, but still a lot of choices!

BTW

I have a couple of tricks I use to help me with permutations and combinations. They both start out the same way—with finding the number of choices you have for each spot you are selecting. The hard part is remembering when to divide by the number of ways to arrange the number of objects you've chosen. I remember when to divide by saying "matters = multiply" and "doesn't = divide." I also use a shortcut to remember what to divide by. Count the spots. Divide by the factorial for that number: Four spots? Divide by 4! Six spots? Divide by 6!

EXERCISES

EXERCISE 10-1

Let's start applying what we've learned about statistics with this group of questions.

1. Which of the following provides the most representative sample of the residents of the state of Iowa?
   a. 10,000 United States citizens, chosen at random
   b. 500 people who currently live in Iowa, chosen at random
   c. 1,000 people who currently live in Iowa, chosen at random
   d. 10 people from each county in Iowa, chosen at random

2. Aaron surveys all the customers in the candy store. He found that 98% of them like to eat candy. He concludes that 98% of the population likes to eat candy. What type of bias is present in this survey?

3. The host of the popular podcast *Trivial* asked listeners to call in and vote for their favorite podcast. Of the callers, 78% said *Trivial* is their favorite podcast. What type of bias is present in this survey?

EXERCISE 10-2

Use the following data set for questions 1–5.

{15, 9, 23, 8, 42, 6, 35, 8}

1. What is the arithmetic mean (average) of the data set?

2. What is the median of the data set?

3. What is the mode of the data set?

4. What is the range of the data set?

5. What would need to be added to the data set to raise the average to 19?

EXERCISE 10-3

Use the following data set for the following questions.

Student	Test 1	Test 2	Test 3	Test 4	Test 5
Bradley	85	87	91	89	92
Audree	96	73	86	90	75

1. What is Bradley's average test score?

2. What is Audree's average test score?

3. What is the mean absolute deviation for Bradley's scores?

4. What is the mean absolute deviation for Audree's scores?

5. Which student has the more consistent performance?

EXERCISE 10-4

Solve the following probability questions. Show your answer as a fraction unless otherwise directed.

1. In a deck of 52 cards, what is the percent probability of drawing one of the hearts?

2. There is a 72% probability of snow on Tuesday. What is the percent probability that it will *not* snow on Tuesday?

3. If a coin is flipped three times, what is the probability of getting tails, then heads, then tails?

4. Rilee has a bag containing three blue marbles and three red marbles. If she draws one, keeps it, and then draws another, what is the probability that both marbles will be blue?

EXERCISE 10-5

Answer the following probability-related questions. Show your answer as a fraction unless otherwise directed.

1. Use the graph here to find the percent probability of a person chosen at random owning either a car or a truck.

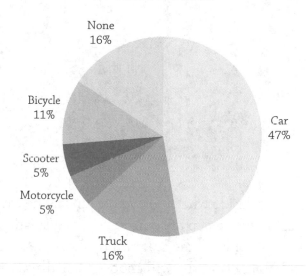

Vehicles Owned

None 16%

Bicycle 11%

Scooter 5%

Motorcycle 5%

Truck 16%

Car 47%

2. Use the following graph to find the probability that it will rain on a randomly chosen day. (Remember that there are 30 days in June and 31 days in July and August.)

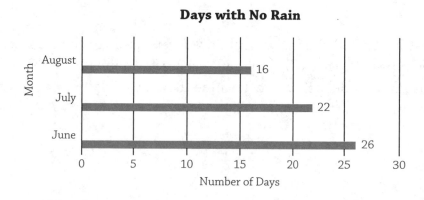

Days with No Rain

August 16

July 22

June 26

Number of Days

Month

3. Use the stem-and-leaf plot to find the probability that a randomly chosen number from the data set will be between 30 and 50.

Stem	Leaf
1	2, 3, 5, 6, 7
2	1, 5, 7, 7, 8
3	2, 4, 4, 5, 6, 8, 9
4	4, 8, 9, 9
5	1, 2, 2, 3, 6, 6
6	1, 7, 8
7	0, 5, 6
8	1, 5, 6, 9, 9
9	2, 3, 6, 7, 8

4. Rhonda is trying to decide what to have for lunch. She can choose a grilled cheese sandwich, a cup of soup, or a Caesar salad, plus an apple, a banana, or a yogurt. Draw a tree diagram to show the possible outcomes. What is the probability that she will choose a salad and a piece of fruit?

EXERCISE 10-6

Read the scenarios presented about probability models and perform each task that follows.

1. In a study of 100 children, the children were asked to choose among three types of toy: a doll, a toy car, or a stuffed animal. Use the probability model below to determine the probability of a child choosing either a stuffed animal or a doll.

Type of Toy	Number of Children	Probability
doll	33	
toy car	26	
stuffed animal	41	
total	**100**	**1**

2. Adam surveyed the 30 students in his class to see what type of home they lived in. He found that 12 students live in a single-family home, 10 live in an apartment, and 8 live in a townhouse. Construct a probability model for this data.

EXERCISE 10-7

Solve the following permutation and combination questions.

1. David is arranging his rock collection on a shelf. He has ten rocks, but only has room for four on the shelf. How many different arrangements of four rocks can he make?

2. Emma is making a seating chart for a formal dinner. Eight people are coming to the dinner, including Emma herself. How many different ways can she arrange the eight people around the table?

3. How many possible groups of three people can be formed from a class of 20 people?

4. A group of six people is choosing a team of two to lead a game. How many different pairs could be formed?

Flashcard App

Geometry Fundamentals

MUST KNOW

 Intersecting lines create pairs of equal, opposing angles.

 Parallel lines cut by a transversal create big angles that are all equal and small angles that are all equal.

eometry is the mathematical study of the size, shape, position, and dimensions of things. Some people do well at geometry quite naturally, and some people find it very difficult. Either way, geometry can be much easier if you remember some basics and use logic to solve for unknowns. Topics in geometry build upon one another, so we need to cover some fundamental concepts before we move on to our study of geometric figures and proportional relationships in the next chapter.

Points, Lines, Line Segments, and Rays

In geometry, a **point** is a specific location in space. It is represented by a dot and is usually labeled with a capital letter. If the point is on the coordinate plane, it can be identified with coordinates, as we saw in Chapter 6.

A **line** is a straight path made up of points. A line goes on forever in both directions, so we can say that a line has infinite length and no real width. We identify lines by naming them with any two points that lie on the line:

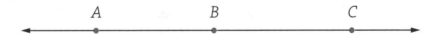

We can call the line above *line AB, AC, BC, BA, CA,* or *CB.* The order in which we write the points does not matter.

EXAMPLE

▶ What are all the possible names for the following line?

▶ Since this line has two points identified, we can call it Line *XY* or Line *YX*.

A **line segment** is a part of a line. It has two ends and is named by the points at which it ends. We write the name of a line segment with a line on top to show that it is a segment rather than a line:

We can call the line segment above $\overline{AB}$ or $\overline{BA}$. The order in which we write the points does not matter. It is the same segment either way.

▶ How many line segments make up this figure?

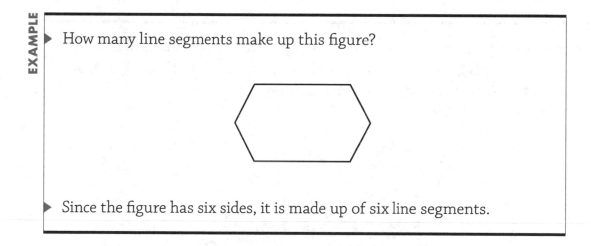

▶ Since the figure has six sides, it is made up of six line segments.

A **ray** is a part of a line that begins at one end point, called a **vertex**, but goes on forever in the other direction. We write the name of a ray by combining the vertex and one other point somewhere on the ray. We use an arrow on top going in the direction of the ray to show that it is a ray rather than a line segment and also to show the direction of the ray:

We can call this ray $\overrightarrow{AB}$ or $\overleftarrow{BA}$. The order in which we write the points does not matter, but the arrow must go in the direction that the line goes on forever. Ray $\overrightarrow{BA}$ is not the same as ray $\overleftarrow{BA}$ because they have different vertexes: Ray $\overrightarrow{BA}$ has vertex B and extends forever in the direction of point A. Ray $\overleftarrow{BA}$ has vertex A and extends forever in the direction of point B. It is more common to write the name of a ray with the arrow going from left to right, so we would usually call the ray shown above $\overrightarrow{AB}$.

EXAMPLE

▶ On the map below, draw a ray that begins at Independence, Missouri, goes through Jefferson City, and then continues east off the right edge of the map. What could we name that ray?

▶ Draw a point for the vertex at Independence. Draw another point at Jefferson City. Draw a line between the two points. Extend the line east to the right edge of the map and draw an arrow to show that the line continues there in that direction.

Tour Route—Missouri

Author: National Park Service

▶ We can name this ray anything we like, with the vertex point first and then the point at Jefferson City, with the arrow on top going from left to right. We could use the city initials and call this ray $\overrightarrow{IJ}$.

Angles

An **angle** is formed by the intersection of two rays with a common vertex. We name angles by a point on one ray, then the vertex, then a point on the other ray. Angle names are usually shown with an angle symbol like ∠ before the letters designating the angle:

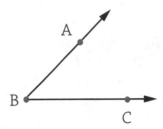

We could call this angle ∠*ABC* or ∠*CBA*. The order of the letters designating the rays does not matter, but the letter representing the vertex must always be in the middle.

Angles are measured in degrees. A line is flat and measures 180°:

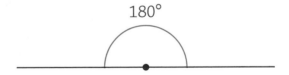

Since a flat line measures 180°, any angles we draw on the line measure less than 180°.

We classify angles based on their degree measurements:

- An angle that measures exactly 90° is called a **right angle**.

- An angle that measures less than 90° is an **acute angle**.

- An angle that measures greater than 90° is called an **obtuse angle**.

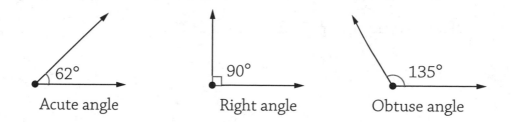

Acute angle   Right angle   Obtuse angle

EXAMPLE

Classify the following angles as right, acute, or obtuse.

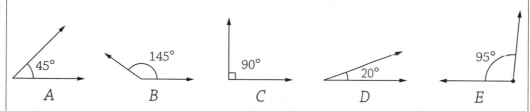

Anything smaller than 90° is acute, so angle *A*, at 45°, and angle *D*, at 20°, are acute.

Anything larger than 90° is obtuse, so angle *B*, at 145°, and angle *E*, at 95°, are obtuse.

An angle that measures exactly 90° is a right angle, so angle *C* is a right angle.

We can combine angles and add their measurements together:

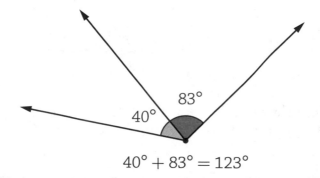

$$40° + 83° = 123°$$

Two 90° angles can be combined to form a 180° line:

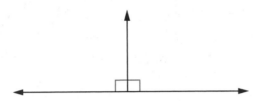

When an obtuse angle is combined with an acute angle to form a 180° line, these are called **supplementary angles**:

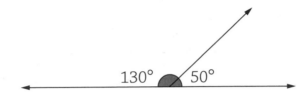

If we know the measurement of one supplementary angle, we can subtract it from 180° to find the measure of the other supplementary angle: 180° − 130° = 50° and 180° − 50° = 130°.

▶ Find the measure of ∠Y.

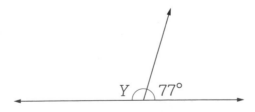

▶ We can see that the two angles form a line, so they must add to 180°. Subtract 77° from 180° to find ∠Y: 180° − 77° = 103°. ∠Y, therefore, equals 103°.

When two acute angles are combined to make a 90° angle, these are called **complementary angles**:

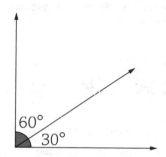

If we know the measurement of one supplementary angle, we can subtract it from 90° to find the measure of the other supplementary angle: $90° - 60° = 30°$ and $90° - 30° = 60°$.

▶ Find the measure of $\angle X$.

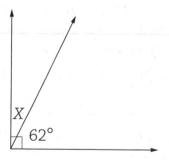

▶ We can see that the two angles form a right angle, so they must add up to 90°. Subtract 62° from 90° to find $\angle X$: $90° - 62° = 28°$. $\angle X$, therefore, equals 28°.

Intersecting and Parallel Lines

When two lines intersect, they form four angles. Each pair of angles opposite each other are called **vertical angles**, and they are equal:

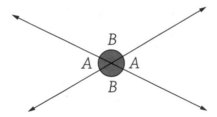

In the preceding drawing, the angles marked A are vertical angles and are equal. The angles marked B are vertical angles and are equal. Angle A does *not* necessarily equal angle B. We can use our knowledge of vertical angles and supplementary angles to find missing angle measures on intersecting lines. As we can see, $\angle A + \angle B = 180°$, $\angle A = \angle A$, and $\angle B = \angle B$.

▶ Find the measures of $\angle X$, $\angle Y$, and $\angle Z$.

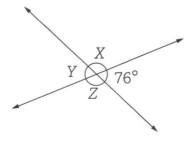

▶ Angle X and $76°$ form a line, so they must add to $180°$. Subtract $76°$ from $180°$ to find $\angle X$: $180° - 76° = 104°$.

▶ Angles X and Z are vertical angles, so they are equal: $\angle Z = 104°$. Angle Y is equal to the $76°$ angle because they are vertical angles: $\angle X = 104°$, $\angle Y = 76°$, and $\angle Z = 104°$.

The lines that form a right angle are called **perpendicular lines**. Any lines that intersect at a right angle are perpendicular:

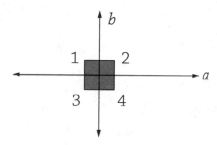

We show that lines are perpendicular with the symbol $\perp$. In the drawing, $a \perp b$. The angles created are all $90°$.

Two straight lines that never intersect at any point are called parallel lines. They look like this:

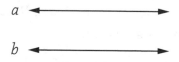

We show that lines are parallel with the symbol $||$. To say that lines a and b are parallel, we can write: $a||b$.

When another line, called a **transversal**, crosses a set of parallel lines, eight angles are formed:

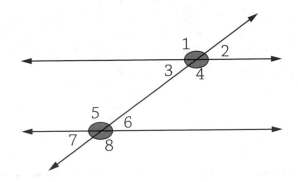

Each set of supplementary angles adds up to 180°:

$$\angle 1 + \angle 2 = 180°$$
$$\angle 3 + \angle 4 = 180°$$
$$\angle 5 + \angle 6 = 180°$$
$$\angle 7 + \angle 8 = 180°$$

$$\angle 1 + \angle 3 = 180°$$
$$\angle 2 + \angle 4 = 180°$$
$$\angle 5 + \angle 7 = 180°$$
$$\angle 6 + \angle 8 = 180°$$

In addition, $\angle 3 + \angle 5 = 180°$ and $\angle 4 + \angle 6 = 180°$. These are called **consecutive interior angles**.

The angles in a pair of vertical angles are equal to each other:

$$\angle 1 = \angle 4$$
$$\angle 2 = \angle 3$$

$$\angle 5 = \angle 8$$
$$\angle 6 = \angle 7$$

If we combine these ideas, we find: $\angle 2 = \angle 3 = \angle 6 = \angle 7$ and $\angle 1 = \angle 4 = \angle 5 = \angle 8$. There are some terms for these angle pairs as well:

- Angles 4/5 and angles 3/6 are known as **alternate interior angles**. Alternate interior angles are equal.

- Angles 1/8 and angles 2/7 are known as **alternate exterior angles**. Alternate exterior angles are equal.

- Angles 1/5 and 4/8, as well as angles 2/6 and 3/7, are known as **corresponding angles**. Corresponding angles are equal.

BTW

All these rules about the angles formed by a line intersecting parallel lines can get confusing. I have an easy way to remember everything:

All the big angles are equal, and all the small angles are equal!

Any big angle plus any small angle equals 180°!

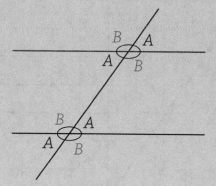

EXERCISES

EXERCISE 11-1

Use the graph below to answer the following questions.

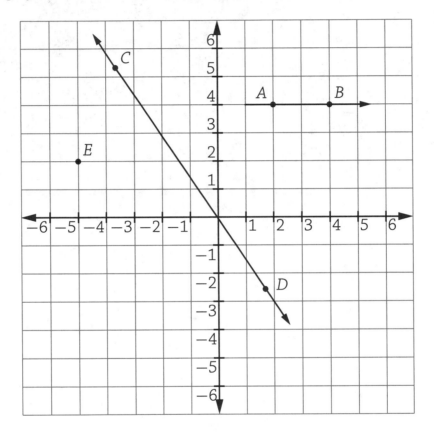

1. What are the coordinates of point *E*?

2. How many points are labeled on the graph?

3. What is the name of a ray shown on the graph?

4. What is the name of a line shown on the graph?

5. What are the names of two line segments shown on the graph?

EXERCISE 11-2

For each question, find the measure of angle x and label it as obtuse, acute, or right.

1.

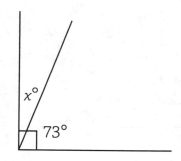

2.

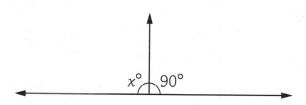

3.

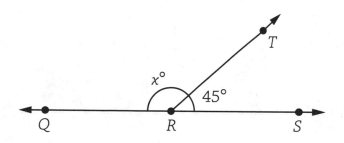

4.

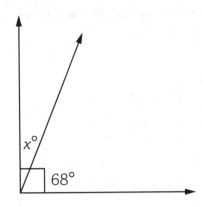

5.

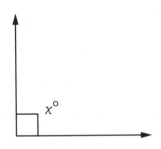

6.

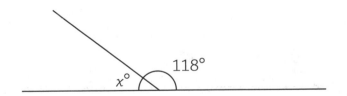

EXERCISE 11-3

Use the following graph to find the measure of each angle or pair of angles.

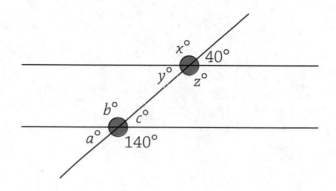

1. $\angle a$

2. $\angle b$

3. $\angle c$

4. $\angle x$

5. $\angle y$

6. $\angle z$

7. $\angle b + \angle y$

8. $\angle c + \angle z$

9. $\angle a + \angle x$

Geometric Figures

MUST KNOW

⚡ Triangles are three-sided shapes. The area of a triangle is given by $A = \frac{1}{2}bh$, and its angles add up to 180°.

⚡ The Pythagorean theorem allows us to find the length of a side of a right triangle: $a^2 + b^2 = c^2$.

⚡ Quadrilaterals are four-sided figures. The angles of all quadrilaterals add up to 360°, while the area is $A = bh$ only for "simple" quadrilaterals.

⚡ A polygon is any two-dimensional shape that is formed by straight sides. Finding the area is easiest by dividing the polygon into shapes such as triangles or rectangles.

⚡ A circle is divided into 360°. We use pi to find its area, $A = \pi r^2$, and its circumference, $C = 2\pi r$, where r is the radius.

277

riangles, quadrilaterals, and circles are all around us. Knowing how to measure these geometric objects and to increase or decrease size proportionally will help us solve many real-world problems. Anytime we build or make something, we are using geometric figures. In this chapter, we will learn to find the perimeter, area, surface area, and volume of common geometric figures. We will also look at proportional relationships in geometry.

Triangles

A **triangle** is a three-sided, two-dimensional figure. There are many different ways to draw a triangle:

Each of the three sides of a triangle can be measured. If all three sides are different lengths, we call that triangle a **scalene triangle**:

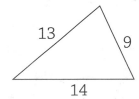

If two of the three sides are equal, we call that triangle an **isosceles triangle**. An isosceles triangle might look like this:

or like this

We often mark the equal sides with those little lines so that other people will know they are equal even without the measurements.

▶ What is the measurement of $\overline{AB}$ below?

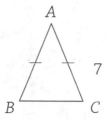

▶ Since this is an isosceles triangle, the marked sides are equal. If $\overline{AC}$ equals 7 then $\overline{AB}$ equals 7, too.

If all three sides of a triangle are equal, we call that triangle an **equilateral triangle**.

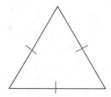

Since a triangle has three sides, it also has three angles formed by those sides. The angles of a triangle *always* add up to 180°. Knowing this allows us to find the measurement of an unknown angle within a triangle if we know the measurements of the other two angles.

▶ What is the measurement of $\angle x$, below?

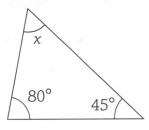

▶ Since we know the angles must add up to 180°, we know: $x + 80° + 45° = 180°$.

▶ We can subtract the angles that are given to us from 180° to find the measurement of angle x: $180° - 80° - 45° = 55°$.

A triangle like the one in the last example is called an **acute triangle** because it has three acute (less than 90°) angles.

A triangle that has an obtuse angle (greater than 90°) is called an **obtuse triangle**:

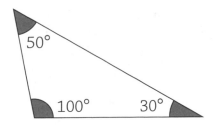

A triangle that has one right angle of 90° is called a **right triangle**. The little box on the triangle's corner indicates that the marked angle is a right angle:

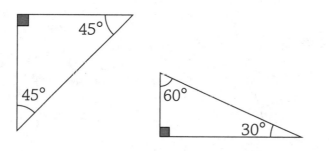

Since all the angles of a triangle must add up to 180°, the other two angles of a right triangle must add up to 90°.

EXAMPLE

▶ What is the measurement of ∠x in the triangle below?

▶ We know that one angle measures 90°, so the sum of the other two angles must also equal 90°: 90° – 34° = 56°.

For an isosceles triangle, finding a missing angle measurement is even easier. Isosceles triangles have two equal sides, so they also have two equal angles opposite the equal sides.

Since the three angles must add up to 180°, if we know one of the angles, we can find all three.

▶ What is the measurement of ∠B in the following triangle? What is the measurement of ∠A?

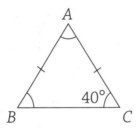

▶ Sides $\overline{AB}$ and $\overline{AC}$ are marked as equal, so ∠B equals ∠C. We know ∠C equals 40°, so ∠B equals 40°, too.

▶ We know that all three angles must add up to 180° so, if ∠C equals 40° and ∠B equals 40°, we can subtract those from 180° to find ∠A: 180° − 40° − 40° = 100°. Therefore, ∠A equals 100°.

For an equilateral triangle, all three angles are equal, just as all three sides are equal. Since the angles in a triangle add up to 180°, this means that each of the three angles is: $180° \div 3 = 60°$.

EXAMPLE

▶ What is the measurement of $\angle ABC$ in the triangle below?

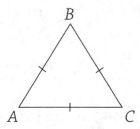

▶ All three sides of this triangle are marked equal. That means it is an equilateral triangle. All the angles are equal; therefore, each angle equals 60°. $\angle ABC$ equals 60°.

The **Pythagorean theorem** can help us find the length of the sides of a right triangle. It states that the square of the **hypotenuse** (the side opposite the right angle) is equal to the sum of the squares of the other two legs:

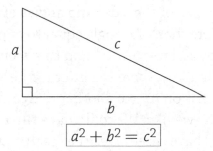

$$a^2 + b^2 = c^2$$

If we know the length of two sides of a right triangle, we can find the length of the third side by using this theorem.

▶ What is the length of side b in the following triangle?

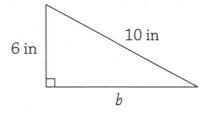

▶ Since this is a right triangle, we can use the Pythagorean theorem to find the missing side.

▶ Let's fill in the values we know for $a^2 + b^2 = c^2$. In this triangle, side a equals 6 and side c equals 10, so we have: $6^2 + b^2 = 10^2$, or $36 + b^2 = 100$. Subtract 36 from both sides to get $b^2 = 64$.

▶ Now we need to take the square root of both sides to solve for b: $\sqrt{b^2} = \sqrt{64}$. Therefore, b equals 8 in.

There is a limit to the size of the sides and angles of a triangle. We cannot make a triangle with more than one angle equal to or greater than 90°. Two 90° angles would already add up to 180°, and that would leave nothing left to form a third angle! Similarly, there is a limit to the lengths of the sides, though that limit is in relation to one another. We could make an equilateral triangle as small or big as we like: We could have three sides that each equal 0.001, or we could have three sides that each equal 1,000,000. For an isosceles triangle, we could have two sides equal to 1,000,000 and one side equal to 0.001, but we cannot have two sides equal to 10 and the third equal to 25. Why not? Try it:

See? Can't do it! With two equal sides of 10, the third side has to be less than 20. There's a rule about this, of course. It's called the **triangle inequality theorem**. It states that the third side of a triangle must be greater than the difference between the other two sides and less than the sum of the other two sides. For an isosceles triangle with two sides of 10, the other side has to be larger than zero but smaller than 20.

EXAMPLE

▶ What is the range of possible values for x in the below triangle?

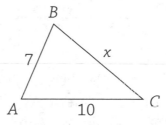

▶ According to the triangle inequality theorem, x has to be greater than the difference between the other two sides: $x > 10 - 7$. That means x is greater than 3.

▶ x also must be less than the sum of the other two sides: $x < 10 + 7$. That means x is less than 17. We can put those results together to form the range of values for x: $17 > x > 3$.

This rule is also important when we are working with the perimeters of triangles. **Perimeter** is the distance around the outside of a figure. For a triangle, this means adding up the lengths of the three sides:

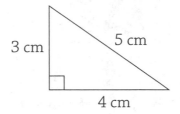

For this triangle, the perimeter equals: 3 cm + 4 cm + 5 cm = 12 cm.
What about this one?

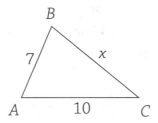

As we saw in the previous example, there is a limit to the value of x. That means there will be a limit to the perimeter as well. To find out what it can be, we first find the range for x. We already found: $17 > x > 3$. Now we add the existing perimeter measurements to that range: $7 + 10 = 17$. We then add 17 to both ends of the range we found for x: $17 + 17 > x > 3 + 17$. The range for the perimeter, p, therefore, is: $34 > p > 20$.

▶ In the triangle that follows, $\overline{AB}$ measures 7.2 cm and $\overline{BC}$ measures 5.4 cm. The perimeter is 19.8 cm. What is the length of $\overline{AC}$?

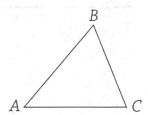

▶ Perimeter (P) is the sum of the sides, and we have two of those measurements. For this triangle, $P = \overline{AB} + \overline{BC} + \overline{AC}$. Let's put in the measurements we have: 19.8 cm $= 7.2$ cm $+ 5.4$ cm $+ \overline{AC}$. We can simplify that to: 19.8 cm $= 12.6$ cm $+ \overline{AC}$.

▶ Now we subtract 12.6 cm from both sides to solve for $\overline{AC}$: 19.8 cm $-$ 12.6 cm $= 12.6$ cm $+ \overline{AC} - 12.6$ cm. Therefore, $\overline{AC}$ equals 7.2 cm.

We have covered angles, sides, and perimeter. What about all that space inside the triangle? We don't want to forget **area**. There is a formula for the area of a triangle: $A = \dfrac{1}{2}bh$, where b is the length of the base, and h is the length of the height:

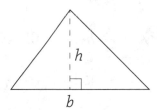

If the base of this triangle is 14, and the height is 8,

$A = \dfrac{1}{2}bh$, so: $\dfrac{1}{2}(14)(8) = (7)(8) = 56$.

▶ What is the area of the below triangle?

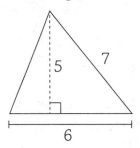

▶ The formula for the area of a triangle is $A = \frac{1}{2}bh$. The base of this triangle is 6, and the height is 5, so: $A = \frac{1}{2}(6)(5) = (3)(5) = 15$.

Since there is a formula for area, we can find any one of the pieces of the formula (area, base, or height) if we have the other two.

▶ If a triangle has an area of 30 and a base of 10, what is the height of the triangle?

▶ The formula for area of a triangle is $A = \frac{1}{2}bh$. Let's fill in the values we know. The area is 30 and the base is 10, so: $30 = \frac{1}{2}(10)(h)$.

▶ Now we can simplify that to: $30 = 5h$. Divide both sides by 5 to solve for h:

$$\frac{30}{5} = \frac{5h}{5}$$
$$6 = h$$

Quadrilaterals

A **quadrilateral** is a four-sided, two-dimensional figure. There are many different types of quadrilaterals. Here are a few of the most common quadrilaterals we use in geometry:

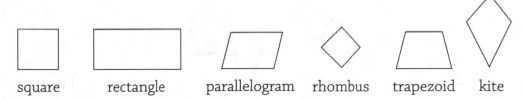

square   rectangle   parallelogram   rhombus   trapezoid   kite

Any four-sided figure is a quadrilateral, even if it is an unfamiliar or irregular shape:

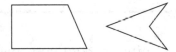

Any quadrilateral can be cut into two triangles:

Since the angles of a triangle add up to 180°, the angles of a quadrilateral add up to 360°.

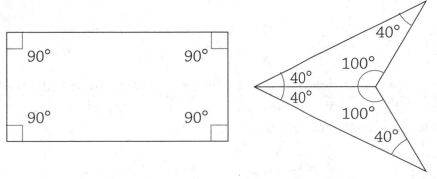

EXAMPLE

▶ What is the measure of $\angle x$ in the below quadrilateral?

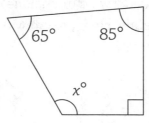

▶ Since all the angles of a quadrilateral add up to 360°, we can add up what we have and subtract it from 360°. Remember that the little box in the lower right corner of the figure shows us that the angle there measures 90°: 65° + 85° + 90°= 240° and 360° − 240° = 120°. Therefore, $\angle x$ equals 120°.

Some quadrilaterals have things in common, and some have features that we can use to find missing sides and angles. Let's look at the defining characteristics of the common quadrilateral shapes:

- **Rectangle**   Opposite sides are equal. Opposite sides are parallel. All angles equal 90°.

- **Square**   *All* sides are equal. Opposite sides are parallel. All angles equal 90°. The only difference between a square and a rectangle is that a square has all four sides equal.

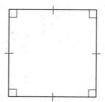

- **Parallelogram**   Opposite sides are equal. Opposite sides are parallel. Opposite angles are equal. The only difference between a parallelogram and a rectangle is that a rectangle has 90° angles.

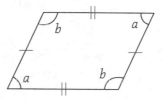

- **Rhombus**   *All* sides are equal. Opposite sides are parallel. Opposite angles are equal. The only difference between a rhombus and a parallelogram is that a rhombus has all four sides equal. The only difference between a rhombus and a square is that a square has 90° angles:

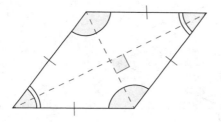

- **Trapezoid**   Only one set of opposite sides is parallel. A **scalene trapezoid** has no equal sides and no equal angles. An **isosceles trapezoid** has two equal sides and two equal angles.

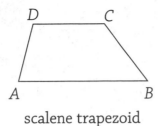

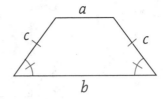

scalene trapezoid   isosceles trapezoid

- **Kite**   There are two sets of equal sides. I like to call them the long sides and the short sides. There is one set of equal angles, and they are the angles between the long and the short sides.

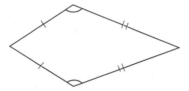

The diagonals of a kite are important too. They are perpendicular to each other (they meet at a right angle), and the long diagonal **bisects** (cuts in half) the short diagonal. In the figure below, $\overline{AC}$ bisects $\overline{BD}$. $\overline{AC}$ also bisects $\angle BAD$ and $\angle BCD$.

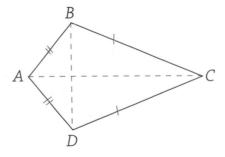

The perimeter of a quadrilateral is the sum of all the sides. For a square or a rhombus, since the sides are all equal, P equals $4s$, where s is the length of a side. For a rectangle or a parallelogram, since opposite sides are equal, we say: $P = 2l + 2w$ (l is length and w is width). A square, for example, with sides of 7.2 has the following perimeter: $P = (4)(7.2) = 28.8$. A rectangle with length 6.5 and width 8.1 has a perimeter of: $P = (2)(6.5) + (2)(8.1) = 13 + 16.2 = 29.2$.

▶ What is the perimeter of the following figure, rounded to the nearest tenth?

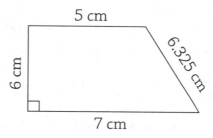

▶ This figure is a trapezoid, and none of the sides are equal. To find the perimeter, we just need to add up the lengths of all the sides: 6 cm + 5 cm + 6.325 cm + 7 cm = 24.325 cm. Rounded to the nearest tenth, the perimeter is 24.3 cm.

Just as we saw with triangles, the perimeter is the distance around the outside, and the area is a measure of the internal space inside the figure.

Do you remember the formula for the area of a triangle? $A = \dfrac{1}{2}bh$. Since all trapezoids can be cut into two triangles, their area is twice the area of a single triangle: $A = bh$. For a rectangle, the base and height are easy to see, so calculating the area is not difficult. Let's look at an example involving that special rectangle, the square.

▶ What is the area of the square below?

3 in

▶ This figure is a square, so all of the sides are equal. *A* equals *bh* for all rectangles, but for a square the base and height are the same; they are both just sides of the square.

▶ Area, then, would be $s \times s$, so we say $A = s^2$. The side of this square is 3 in, so: $A = (3 \text{ in})^2 = 9 \text{ in}^2$. The area is 9 in^2.

▶ The unit *inch* is squared just as the number 3 is squared, so we write 9 in^2.

Finding area is a little more complicated with other types of quadrilaterals. Parallelograms and rhombuses are like slanted rectangles, so all we have to do to find their area is be sure we are measuring the actual height rather than the slanted side.

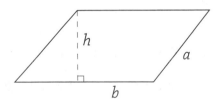

In this figure, *b* is the base, and *h* is the height. *A* is the side, not the height.

EXAMPLE

▶ What is the area of the below rhombus?

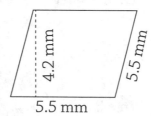

4.2 mm

5.5 mm

5.5 mm

▶ Since this figure is a rhombus, all of the sides are equal. *A* equals *bh*, but we need to be careful that we use 4.2 mm as the height rather than 5.5 mm (which is the side).

▶ The base of this rhombus is 5.5 mm, and the height is 4.2 mm, so:
$A = 5.5 \text{ mm} \times 4.2 \text{ mm} = 23.1 \text{ mm}^2$.

Trapezoids are a bit more difficult. Since the base and the top side of a trapezoid are not the same length, we can't use the formula $A = bh$. What we do instead is find the *average* of the parallel sides (the base and the top side) and then multiply that times the height.

EXAMPLE

▶ What is the area of the following trapezoid?

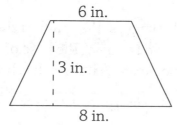

6 in.

3 in.

8 in.

▶ *A* equals *bh* but, since this figure is a trapezoid, we need to average the lengths of the top and bottom sides to get a measurement to use for the base:

$$\frac{6\,\text{in} + 8\,\text{in}}{2} = \frac{14\,\text{in}}{2} = 7\,\text{in}$$

▶ The height is 3 in, so: $A = 7\,\text{in} \times 3\,\text{in} = 21\,\text{in}^2$.

Finding the area of a kite is quite different. We can turn a quadrilateral any way we like to make any of the four sides the base, but with a kite there doesn't seem to be an obvious base. The formula for the area of a kite is $A = \dfrac{(d_1)(d_2)}{2}$, where d_1 and d_2 are the lengths of the two diagonals. Why are we using diagonals to find area? Since there isn't an obvious base, we break the kite into two triangles:

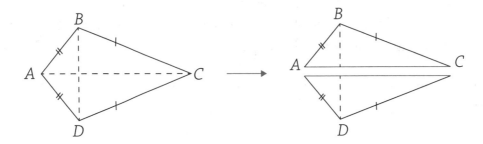

Now we have two equal triangles. We can find the area of one triangle and double it. Let's use the top triangle. The base of that triangle is the long diagonal (d_1) of the kite. The height is half of the short diagonal (d_2) of the kite. The area of a triangle is $\dfrac{1}{2}bh$, so we get: $A = \left(\dfrac{1}{2}\right)(d_1)\left(\dfrac{1}{2}d_2\right)$. We then double that area to find the area of the whole kite, so the area of the kite is:

$A = (2)\left(\dfrac{1}{2}\right)(d_1)\left(\dfrac{1}{2}d_2\right)$. We can simplify that to: $A = (d_1)\left(\dfrac{1}{2}d_2\right)$,

or $A = \dfrac{(d_1)(d_2)}{2}$.

For this kite, let's say $\overline{AC}$ equals 6 and $\overline{BD}$ equals 4. The area of the top triangle then would be A equals $\dfrac{1}{2}bh$, so: $A = \left(\dfrac{1}{2}\right)(6)(2) = 6$. We double that to find the area of the whole kite: $6 \times 2 = 12$. Using the formula $A = \dfrac{(d_1)(d_2)}{2}$, we have: $A = \dfrac{(6)(4)}{2} = \dfrac{24}{2} = 12$. Same area either way!

▶ What is the area of the kite below if $\overline{KM}$ equals 5.2 meters and $\overline{JL}$ equals 3.9 meters?

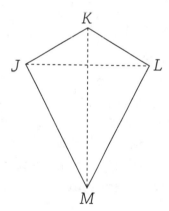

▶ Using the formula $A = \dfrac{(d_1)(d_2)}{2}$, we have:

$$A = \dfrac{(5.2\,\text{m})(3.9\,\text{m})}{2} = \dfrac{20.28\,\text{m}^2}{2} = 10.14\,\text{m}^2$$

Polygons

A **polygon** is, simply, a two-dimensional shape with straight sides. Every figure we have looked at so far is a polygon. There are two types of polygons: regular and irregular. A **regular polygon** has all sides equal and all angles equal. Here are some examples of regular polygons:

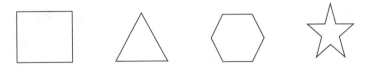

Notice that these include squares and equilateral triangles, which we have already discussed.

An **irregular polygon** does not have all sides equal and all angles equal, although some of the angles and sides may be equal. Here are some examples of irregular polygons:

Notice that these include rectangles, triangles that are not equilateral, parallelograms, rhombuses, and trapezoids, which we have already discussed. We know how to deal with those figures, but what about the unfamiliar ones? How can we find the area or perimeter of shapes like these:

The key here is to break these figures into more familiar shapes, as we did with the kite:

can be split into two rectangles, like this:

If we know the length and width of both, we can find the perimeter or area of both and then add them together.

▶ What is the perimeter of the figure below? What is its area?

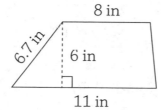

▶ If we think of the dotted line showing the height of the figure as a solid line dividing the figure into a triangle and a rectangle, we can find the perimeter and area of each one and then add them together to find the total perimeter and area.

▶ The top of the rectangle is marked as 8 in. The bottom of the rectangle must also be 8 in. The left side of the rectangle (the dividing line of the figure) is marked as 6 in, so the right side of the rectangle must also be 6 in.

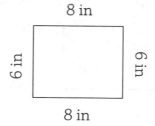

► The triangle has a base of 3 in, since the original base was 11 in and we subtracted 8 in for the rectangle. The height of the triangle is 6 in.

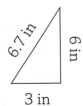

► Now we can put everything together. The perimeter of the figure is 6.7 in + 8 in + 11 in + 6 in (for the right side only; do not include the 6 in marked for the dotted line, since that is not the on the outside of the figure. P = 31.7 in.

► For the area, we need the area of the rectangle, $A = lw$, plus the area of the triangle, $A = \frac{1}{2}bh$. The width of the rectangle is 6 in, and the length is 8 in, so: $A = (6 \text{ in})(8 \text{ in}) = 48 \text{ in}^2$.

► The base of the triangle is 3 in, and the height is 6 in, so: $A = \frac{1}{2}(3 \text{ in})(6 \text{ in}) = 9 \text{ in}^2$. Add those together to find the total area: $48 \text{ in}^2 + 9 \text{ in}^2 = 57 \text{ in}^2$.

We can use this dividing procedure to find the perimeter and area of lots of different figures. Regular pentagons, hexagons, and octagons can all be divided into equal triangles. Since a regular pentagon has five equal sides, we can divide it like this:

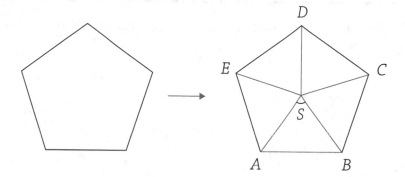

If one side measures 2 feet, then all five sides of the pentagon measure 2 feet, and the perimeter will be: $5s = 5(2 \text{ ft}) = 10 \text{ ft}$.

When we divide the pentagon into five equal triangles, we are also dividing the center into five equal angles. Since a circle measures 360°, when we divide 360° by five, we find that each central angle equals 72°:

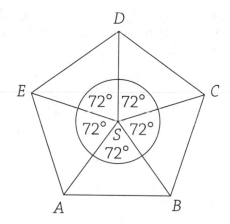

Each of the five triangles formed has two equal sides. For triangle ASB, $\overline{SA}$ equals $\overline{SB}$. Since each triangle is an isosceles, we also know that $\angle SAB$ equals $\angle ABS$. We can subtract 72° from the total 180° in a triangle to get: $180° - 72° = 108°$. Divide that result in half to find the measure of $\angle SAB$ and $\angle ABS$: $108° \div 2° = 54°$. Each of the five isosceles triangles has two 54° angles and one 72° angle.

▶ What is the perimeter of the following regular hexagon? What is the measure of ∠F?

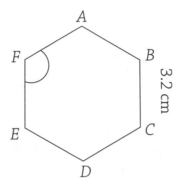

▶ This is a regular hexagon, which means all six sides are equal. If one side measures 3.2 centimeters, then the perimeter is: $6s = 6(3.2 \text{ cm})$ $= 19.2$ cm.

▶ To find ∠F, we can cut the hexagon into six equal triangles, like this:

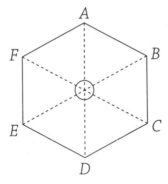

▶ The six central angles are equal and must add up to 360°: $360° \div 6 = 60°$. That means the sum of the other two angles of each triangle is: $180° − 60° = 120°$ and $120° \div 2 = 60°$. Therefore, each of the six triangles we made is an equilateral triangle.

▶ All three angles are 60° and all three sides are equal. ∠F is made up of two of these 60° angles, so ∠F = 120°.

Circles

You are certainly familiar with what a circle looks like, but let's make sure we know all the related terms:

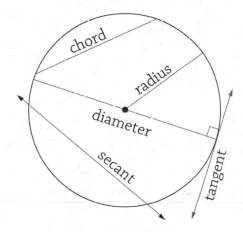

The **diameter** of a circle is the distance straight across the middle. A **radius** is any line drawn from the center of a circle to the outside edge; it is half the diameter. A **secant** is a line that goes through a circle but does not touch the center. A **chord** is any line segment within a circle other than the diameter; the part of a secant that is inside a circle is a chord. A **tangent** is a line that touches the outside of a circle at just one spot.

The degree measure of a circle is 360°. A circle cut into fourths makes four right angles, and 4 × 90° = 360°. A semicircle has 180° because the diameter is a straight line, as we can see on the next page.

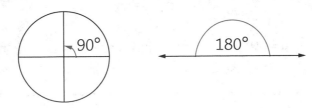

EXAMPLE

▶ What is the measure of ∠x, below?

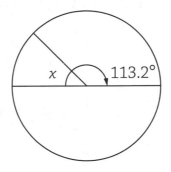

▶ The diameter forms a 180° angle and the marked obtuse angle is 133.2°, so we can subtract that from 180° to find x: 180° − 133.2° = 46.8°.

The perimeter, or distance around the outside of the circle, is called the **circumference**. The formula for finding circumference is $C = \pi d$, where d is the length of the diameter, or $C = 2\pi r$, where r is the length of a radius.

EXAMPLE

▶ What is the circumference of the following circle?

▶ The formula for circumference is $C = \pi d$, or $C = 2\pi r$. We are given the length of a radius, so let's use $C = 2\pi r$: $C = 2\pi(9 \text{ in}) = 18\pi$ in.

▶ The circumference of this circle is 18π in. If we use 3.14 for π, we can multiply: $18 \times 3.14 \approx 56.52$ in.

BTW

*We usually leave π, which is called **pi**, as a symbol.*

The formula for finding the area of a circle is $A = \pi r^2$, where r is the length of a radius. In the circle we just looked at, the radius is 9 in, so the area is: $A = \pi(9 \text{ in})^2 = 81\pi \text{ in}^2$. Remember to square the units as well as the radius.

EXAMPLE

▶ What is area of the circle below?

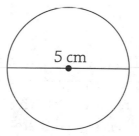

5 cm

▶ The formula for the area of a circle is $A = \pi r^2$, but we are given the measurement of the diameter.

▶ We need to cut the diameter in half to find the radius. The radius is 2.5 centimeters, so: $A = \pi(2.5 \text{ cm})^2 = 6.25\pi \text{ cm}^2$.

Surface Area

All the figures we have looked at so far have been two-dimensional. What about three-dimensional figures such as cubes, boxes, pyramids, and spheres? We call the space inside these figures **volume**, and the space all around the outside is called **surface area**. Surface area is found by adding up the areas of all the faces of the object. In other words, it's the area of all the surfaces.

A **cube** is like a square but in three dimensions. Each face (side) of a cube is a square, and a cube has six faces. Think of a cube like a square box that we can open and unfold until it is flat:

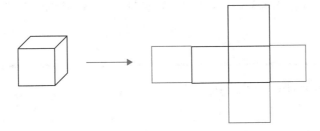

To find the surface area of a cube, we need to find the areas of all six faces and then add them together. Since a square has equal sides, we found that the area of a square is s^2, where s is the length of one side of the square. If the area of one face of the cube is s^2, and there are six equal faces, then the surface area of a cube is: $S_A = 6s^2$. For a cube with a side of 3, the surface area is: $(6)(3^2) = (6)(9) = 54 \text{ units}^2$.

EXAMPLE

▶ What is the surface area of the following cube?

$3\frac{1}{2}$ cm

> The area of each face of the cube is $A = s^2$, and there are six faces, so the surface area is $S_A = 6s^2$.

> Each side measures $3\frac{1}{2}$ cm, but let's convert the fraction to a decimal for easier calculations: $S_A = 6 \times (3.5 \text{ cm})^2 = 6 \times 12.25 \text{ cm}^2 = 73.5 \text{ cm}^2$.

To find the surface area of a rectangular box, which in fancy geometry language is called a **right rectangular prism**, we find the area of each face and add them together. If we unfolded the box, it would look like this:

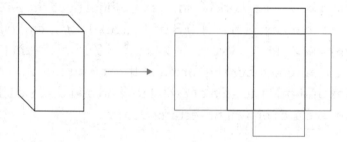

As you can see, the box has three sets of two equal sides. I call them top/bottom, front/back, and side/side. For a box that looks like this one, each pair has a different area, so we will have to find three different areas and then remember that there are two equal sides for each kind of area:

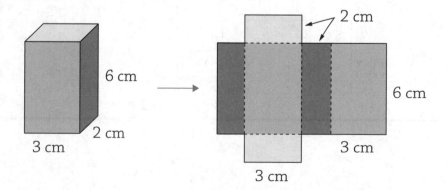

The top/bottom sides each measure 2 cm by 3 cm, so the area is 6 cm² for each of them. The front/back sides each measure 3 cm by 6 cm, so the area is 18 cm² for each of them. The side/side sides each measure 2 cm by 6 cm, so the area is 12 cm² for each of them. Now let's add everything up:

top: 6 cm²   front: 18 cm²   side: 12 cm²
bottom: 6 cm²   back: 18 cm²   side: 12 cm²

→ 6 cm² + 6 cm² + 18 cm² + 18 cm² + 12 cm² + 12 cm² = 72 cm²

It's easier with a cube, right?

There is a formula for surface area of a right rectangular prism: $2lw + 2lh + 2wh$, where l is the length, w is the width, and h is the height. If we used the formula on the box we just worked on, we would just put the length, width, and height into the formula: $S_A = (2)(3 \text{ cm})(2 \text{ cm}) + (2)(3 \text{ cm})(6 \text{ cm}) + (2)(2 \text{ cm})(6 \text{ cm}) = 12 \text{ cm}^2 + 36 \text{ cm}^2 + 24 \text{ cm}^2 = 72 \text{ cm}^2$. Is that easier? Maybe—if you can remember the formula! If not, you can always mentally unfold that box and find the area of each face and add them up. Now let's try finding the surface area of a right rectangular prism.

▶ What is area of the right rectangular prism below?

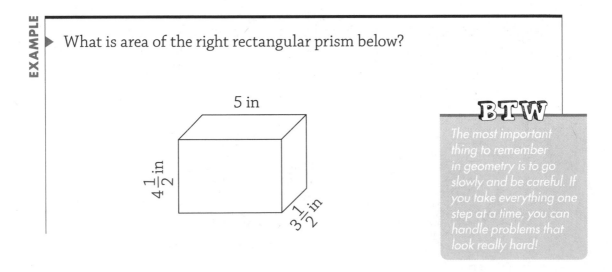

5 in

4½ in

3½ in

BTW

The most important thing to remember in geometry is to go slowly and be careful. If you take everything one step at a time, you can handle problems that look really hard!

▶ First, find the area of each face. Let's use decimals rather than fractions. The top/bottom sides each measure 3.5 in by 5 in, so the area is 17.5 in² for each of them. The front/back sides each measure 5 in by 4.5 in, so the area is 22.5 in² for each of them. The side/side sides each measure 3.5 in by 4.5 in, so the area is 15.75 in² for each of them.

▶ Now let's add everything up: 17.5 in² + 17.5 in² + 22.5 in² + 22.5 in² + 15.75 in² + 15.75 in² = 111.5 in². If we use the formula, we have:
$S_A = (2)(5 \text{ in})(3.5 \text{ in}) + (2)(5 \text{ in})(4.5 \text{ in}) + (2)(3.5 \text{ in})(4.5 \text{ in}) = 35 \text{ in}^2 + 45 \text{ in}^2 + 31.5 \text{ in}^2 = 111.5 \text{ in}^2$.

What about something like this?

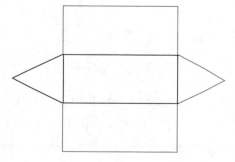

This is called a **triangular prism** because the ends are triangles. If we unfold this prism, we get three rectangles and two triangles:

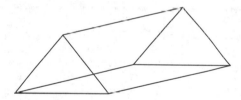

To find the surface area, we find the area of each rectangle and the area of each triangle (the area of the triangles will be the same, so we just need to find the area of one and double it) and then add those up.

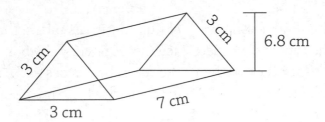

Each of the triangles has a base of 3 cm and a height of 6.8 cm: $A = \frac{1}{2}bh$, so $A = \frac{1}{2}(3\text{ cm})(6.8\text{ cm}) = 10.2\text{ cm}^2$. The bottom rectangle is 3 cm by 7 cm, so its area is 21 cm². The left side rectangle is also 3 cm by 7 cm, so its area is 21 cm². The right-side rectangle is also 3 cm by 7 cm, so its area is 21 cm². Add all the areas up: $10.2\text{ cm}^2 + 10.2\text{ cm}^2 + 21\text{ cm}^2 + 21\text{ cm}^2 + 21\text{ cm}^2 = 83.4\text{ cm}^2$. Now let's try one together.

EXAMPLE

▶ Find the surface area of the following triangular prism.

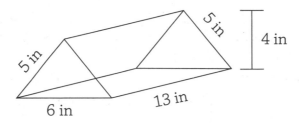

▶ Each of the triangles has a base of 6 in and a height of 4 in, so:

$A = \frac{1}{2}(6\text{ in})(4\text{ in}) = 12\text{ in}^2$. The bottom rectangle is 6 in by 13 in, so its area is 78 in². The left-side rectangle is 5 in by 13 in, so its area is 65 in². The right-side rectangle is also 5 in by 13 in, so its area is 65 in².

> Add up all the areas: $12 \text{ in}^2 + 12 \text{ in}^2 + 78 \text{ in}^2 + 65 \text{ in}^2 + 65 \text{ in}^2 =$ 232 in^2.

There is one more very common figure we need to look at for surface area: the **cylinder**.

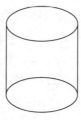

If we "unfold" a cylinder, it looks like two equal circles and a rectangle. The width of the rectangle is the same as the circumference of the circle:

To find the surface area of a cylinder, we need to find the area of the rectangle plus the area of the two equal circles. We will need to have the radius of the circle to find its area and the height of the cylinder to find the area of the rectangle.

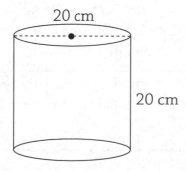

20 cm

20 cm

First, let's find the area of the rectangle. It measures 20 cm by whatever the circumference of the circle is, so we need to find that. The diameter is 20 cm, and C equals πd, so: C equals 20π cm. The rectangle measures 20 cm by 20π cm, so its area is 400π cm^2. Now let's find the area of the circles. Each circle has a diameter of 20 cm, so the radius is 10 cm: $A = \pi r^2$, so $A = \pi \times (10 \text{ cm})^2 = 100\pi$ cm^2. Add the area of both circles to the area of the rectangle to get the total surface area: $S_A = 100\pi$ cm$^2 + 100\pi$ cm$^2 + 400\pi$ cm$^2 = 600\pi$ cm^2.

If we need to calculate using pi rather than leaving it as a symbol, we can estimate pi as 3.14, so 600π cm$^2 = 1{,}884$ cm^2. There is a formula for surface area of a cylinder too: $S_A = 2\pi r^2 + 2\pi rh$. If you look closely at the formula, you can see that it is exactly what we did: two times the area of the circle (πr^2) plus the product of the circumference $(2\pi r)$ and the height, which is the area of the rectangle. Let's work with an example.

> What is the surface area of the below cylinder?

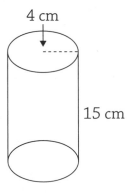

4 cm

15 cm

> Let's use the formula $S_A = 2\pi r^2 + 2\pi rh$. The radius is 4 cm, and the height is 15 cm, so we have: $S_A = 2\pi(4 \text{ cm})^2 + 2\pi(4 \text{ cm})(15 \text{ cm}) = 32\pi$ cm$^2 + 120\pi$ cm$^2 = 152\pi$ cm^2.

> If you can't remember the formula, you can always mentally unfold the cylinder and see that it is a rectangle and two circles. Find each area and add them together.

Volume

What about the space inside a three-dimensional figure? How much can we stuff into a box? How much soda can we put in a cylindrical glass? To find out, we need to calculate the volume that the object can hold. For things like boxes and cylinders, we can think of volume as the area of the base of the object times the height of the object.

Naturally, we have formulas for each type of figure, but if you remember that volume is the area of the base times the height, you can find the volume of anything that doesn't change shape as the height increases. Shapes for which the height changes include pyramids, cones, etc. They require formulas, which will usually be given to you, so there is no need to memorize those formulas.

Let's start with a cube. The area of the base is s^2, and the height is the same as the side, so we are multiplying: $s^2 \times s = s^3$. This is the same as multiplying all three dimensions (length, width, and height) together, but for a square they are all equal, so we just say: $V = s^3$. If we have a square with a side of 4 inches, the volume is: $V = (4 \text{ in})^3 = 64 \text{ in}^3$. Remember to cube the units as well as the number.

EXAMPLE

▶ What is volume of the below cube, rounded to the nearest whole number?

> V equals s^3, and each side measures $3\frac{1}{2}$ cm, but let's convert that to a decimal for easier calculations. Each side measures 3.5 cm:
> $V = (3.5 \text{ cm})^3 = 42.875 \text{ cm}^3$.

> When we round that to the nearest whole number, we have 43 cm^3.

For a right rectangular prism, the area of the base is l times w, so the area of the base times the height is: $l \times w \times h$. If we have a box that measures 12 in by 24 in by 4 in, the volume is: $V = (12 \text{ in})(24 \text{ in})(4 \text{ in}) = 1{,}152 \text{ in}^3$.

EXAMPLE

> What is the volume of the following right rectangular prism?

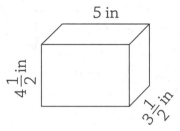

> V equals l by w by h. Let's use decimals rather than fractions. The length is 5 in, the width is 3.5 in, and the height is 4.5 in: $V = (5 \text{ in})(3.5 \text{ in})(4.5 \text{ in}) = 78.75 \text{ in}^3$.

We can follow a similar procedure to find the volume of a triangular prism. We find the area of the triangular base and multiply that by the height (length) of the prism. The formula for the volume of a triangular prism is $V = \dfrac{1}{2}bhl$, where l is the length of the prism, b is the base of the triangle, and h is the height of the triangle:

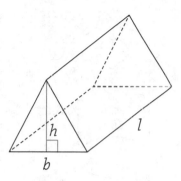

EXAMPLE

▶ What is the volume of the triangular prism below?

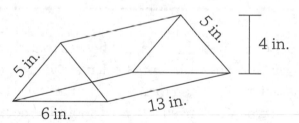

▶ V equals $\dfrac{1}{2}bhl$. The base of the triangle is 6 in, and the height is 4 in. The length of the prism is 13 in. V equals $\dfrac{1}{2}bhl$ so:

$$V = \frac{1}{2}\left(6\,\text{in}\right)\left(4\,\text{in}\right)\left(13\,\text{in}\right) = 156\,\text{in}^3.$$

For a cylinder, the area of the base is the area of a circle: $A = \pi r^2$. Multiply that by the height to get the volume of the cylinder. As you may have guessed, the formula is $V = \pi r^2 h$. If we have a cylinder with a radius of 2 cm and a height of 12 cm, the volume is: $V = \pi(2\text{ cm})^2 (12\text{ cm}) = (4\pi\text{ cm}^2)(12\text{ cm}) = 48\pi\text{ cm}^3$.

What is volume of the following cylinder?

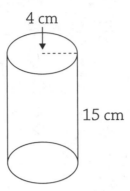

4 cm

15 cm

V equals $\pi r^2 h$. The radius is 4 cm and the height is 15 cm:
$V = \pi(4 \text{ cm})^2(15 \text{ cm}) = (16\pi \text{ cm}^2)(15 \text{ cm}) = 240\pi \text{ cm}^3$.

Now let's look at some three-dimensional figures that change shape as their height changes. For these, we cannot calculate volume as the area of the base times the height. We need to use a formula. These formulas are more complicated, so you won't need to worry about memorizing them. You can always look them up if you need them.

A **sphere** is a ball, but if you take a cross section of the sphere, you will find a circular surface.

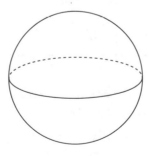

The formula for volume of a sphere is $V = \dfrac{4}{3}\pi r^3$, where r is the radius of the sphere.

EXAMPLE

▶ What is volume of a hemisphere if $r = 6$ cm?

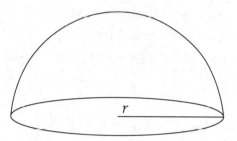

▶ Since a hemisphere is a sphere cut in half, we can find the volume of the whole sphere and then cut it in half: $V = \dfrac{4}{3}\pi r^3$. The radius is 6 centimeters, so:

$$V = \frac{4}{3}\pi\left(6\,\text{cm}\right)^3 = \frac{4}{3}\pi\left(216\,\text{cm}^3\right) = 288\pi\,\text{cm}^3$$

▶ To find the volume of the hemisphere, we just need to cut that in half: $288\pi\,\text{cm}^3 \div 2 = 144\pi\,\text{cm}^3$.

A **cone** looks like this:

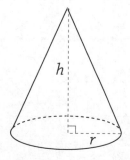

The formula for the volume of a cone is $V = \frac{1}{3}\pi r^2 h$, where r is the radius of the circular base, and h is the height of the cone (through the center).

▶ Which has a greater volume: a cone with height 6 and diameter 4 or a cone with height 4 and diameter 6?

▶ Here we need to compare the volumes of two different cones. Let's name them for convenience. Cone A has height 6 and diameter 4, which means the radius is 2. Cone B has height 4 and diameter 6, which means the radius is 3.

▶ The formula for the volume of a cone is $V = \frac{1}{3}\pi r^2 h$. Cone A's volume is:

$$V = \frac{1}{3}\pi(2^2)(6) = \frac{1}{3}\pi(4)(6) = \frac{1}{3}\pi(24) = 8\pi$$

▶ Cone B's volume is:

$$V = \frac{1}{3}\pi(3^2)(4) = \frac{1}{3}\pi(9)(4) = \frac{1}{3}\pi(36) = 12\pi$$

▶ Cone B holds a greater volume, so the greater volume is a cone with height 4 and diameter 6.

A **pyramid** looks like this:

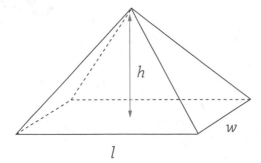

The formula for the volume of a pyramid is $V = \dfrac{lwh}{3}$, where l and w are the length and width of the base, and h is the height of the pyramid (through the center).

▶ What is the volume of the pyramid below?

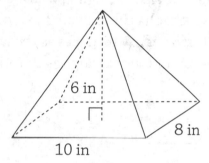

6 in

8 in

10 in

▶ The pyramid has a 10-in by 8-in base and a height of 6 in. $V = \dfrac{lwh}{3}$, so:

$$V = \frac{(10\,\text{in})(8\,\text{in})(6\,\text{in})}{3} = \frac{480\,\text{in}^3}{3} = 160\,\text{in}^3$$

IRL Volume is very important. Volume tells us how much something holds: cans, boxes, swimming pools, a gas tank, etc.

Geometric Transformations

In geometry, proportional relationships are referred to as **scaling**. Scaling means making a figure larger or smaller but still in the same proportions as the original. We call these **similar** figures. If we take a rectangle that is 2 cm by 4 cm and want to make it bigger, we can use our knowledge of proportions to help us. If we want a rectangle that is twice as large as the original, we will need to double both the length and the width:

$$\frac{2\,cm}{4\,cm} = \frac{4\,cm}{8\,cm}$$

The same is true for any other figure. If we want to increase or decrease the size, we must do the same to each dimension of the figure. We call the process of increasing or decreasing the size of a figure, while retaining its proportions, **dilation**. Here is a triangle that has been dilated to twice its original size. We say that there is a **scale factor** of 2, which means the ratio of the new triangle to the original triangle is 2:1, or twice as large as the original triangle:

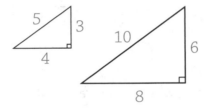

 IRL Architects, engineers, artists, and designers use scale to help them visualize their work. A scale model of a building or the floor plan of a house are just a couple of examples of how we use scale.

EXAMPLE

The floor plan for a new house has a scale of $\frac{1}{4}$ inch $= 1$ foot. The floor plan for the living room is shown here. What are the actual dimensions of the living room?

▶ Let's set up a proportion for the length. Since the dimensions of the drawing are given in decimal form, let's use decimals. We can use x for the unknown length.

$$\frac{0.25\,\text{in}}{1\,\text{ft}} = \frac{4.25\,\text{in}}{x\,\text{ft}}$$

▶ Do you remember how to solve a proportion? Cross multiply.

$$(0.25\text{ in}) (x\text{ ft}) = (1)(4.25\text{ in})$$
$$(0.25\text{ in}) (x\text{ ft}) = (4.25\text{ in})$$

▶ Divide both sides by 0.25 in to solve for x.

$$\frac{(0.25\text{ in})(x\text{ ft})}{0.25\text{ in}} = \frac{4.25\text{ in}}{0.25\text{ in}}$$
$$x = 17\,\text{ft}$$

▶ The length of the actual living room is 17 feet. Now let's do the same for the width, using w for the unknown width.

$$\frac{0.25\,\text{in}}{1\,\text{ft}} = \frac{3.125\,\text{in}}{w\,\text{ft}}$$
$$\left(0.25\text{ in}\right)\left(w\text{ ft}\right) = (1)\left(3.125\text{ in}\right)$$
$$\left(0.25\text{ in}\right)\left(w\text{ ft}\right) = \left(3.125\text{ in}\right)$$

▶ Divide both sides by 0.25 in to solve for w.

$$\frac{(0.25\text{ in})(w\text{ ft})}{0.25\text{ in}} = \frac{3.125\text{ in}}{0.25\text{ in}}$$
$$w = 12.5\,\text{ft}$$

▶ The width of the actual living room is 12.5 feet, so the actual dimensions of the living room are 17 ft by 12.5 ft.

A geometric **transformation** is a change that can be made to a figure by turning it, flipping it, sliding it, or resizing it. We have been looking at dilation, which is one type of transformation that produces a similar figure that is proportional to the original. The other types of transformations produce figures that are **congruent** rather than similar. Congruent figures have the same shape and size as the original:

- **Rotation** is a transformation in which we turn the figure around. Here is a triangle that has been rotated to the right. The new triangle is congruent to the original.

- **Point symmetry** is produced when we rotate a figure 180° around a central point. The curve on this graph has been rotated 180° around the origin.

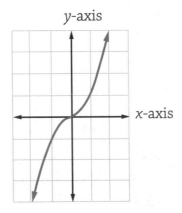

- **Translation** is a transformation in which we slide the figure in some direction. Here is a triangle that has been translated slightly up and to the right on the coordinate plane. The new triangle is congruent to the original.

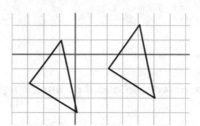

- **Reflection** is a transformation in which we flip the figure over, creating a mirror image. Here is a triangle that has been reflected across the *x*-axis. The new triangle is congruent to the original.

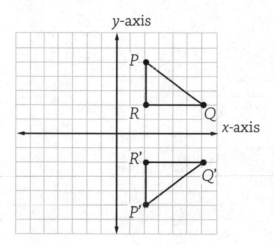

Line symmetry is produced when we reflect a figure. A line of symmetry occurs when a figure can be folded in half along a line and the two halves match up perfectly. Here are some figures that have lines of symmetry:

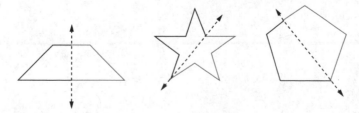

EXERCISES

EXERCISE 12-1

Let's start applying what we've learned about geometric figures to the following questions.

1. Label each triangle as scalene, isosceles, right, or equilateral.

   a.

   b.

   c.

   d.

   e.

   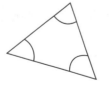

2. Label each angle as acute, obtuse, or right.

a.

b.

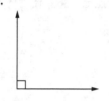

c.

d.

3. What is the measure of angle *x* below?

a.

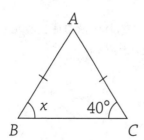

b.

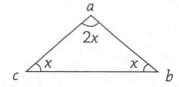

c.

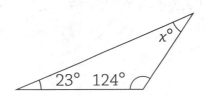

4. In the following two triangles, what is the measure of $\overline{AC}$ if $\overline{AB} = 4$ and $\overline{BC} = 4$? What is the area of triangle b?

a.

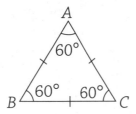

b.

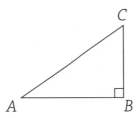

5. What is the range of possible values for $\overline{AB}$? What is the range of possible values for the perimeter of the triangle?

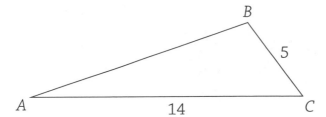

6. What is the area of the triangle? What is the perimeter?

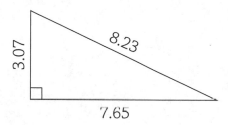

7. What is the perimeter of the triangle?

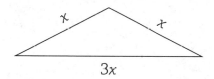

EXERCISE 12-2

Follow the instructions for each group of questions about polygons.

1. Name each quadrilateral. What is the measure of $\angle x$ in each?

a.

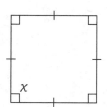

b.

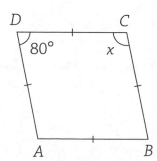

c.

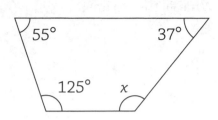

d.

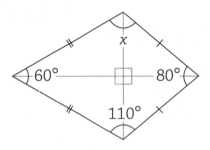

e.

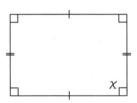

f.

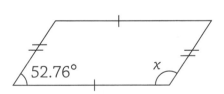

2. What is the area of each figure?

a.

b.

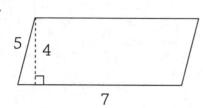

c.

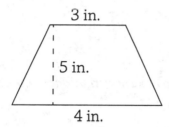

d.

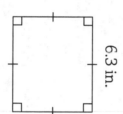

e.

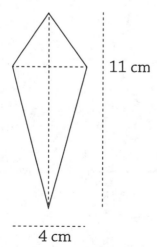

f.

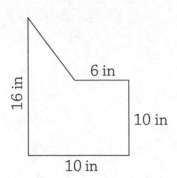

EXERCISE 12-3

Now, let's apply what we've learned about circles to answering a few questions.

1. What is the measure of $\angle x$?

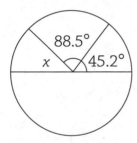

2. What is the circumference of the circle?

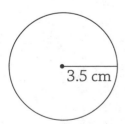

3. What is the area of the circle?

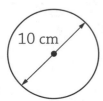

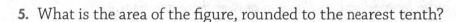

4. If a circle has an area of 9π, what is the length of the diameter?

5. What is the area of the figure, rounded to the nearest tenth?

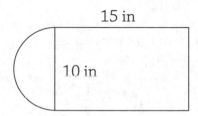

15 in

10 in

EXERCISE 12-4

Answer these questions about surface area.

1. What is the surface area of a cube with sides of 3.1 cm?

2. What is the surface area of a right rectangular prism that measures 4 in by 5 in by 7 in?

3. What is the surface area of the below triangular prism?

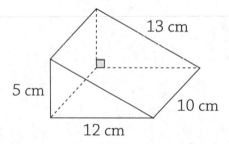

13 cm

5 cm

12 cm

10 cm

4. What is the surface area of the following cylinder?

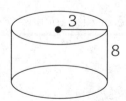

3

8

EXERCISE 12-5

Apply what you've learned about volume to answer these questions.

1. What is the volume of a cube with sides of 2.4 in.?

2. What is the volume of a right rectangular prism that measures
    3 millimeters by 6.5 millimeters by 2 millimeters?

3. What is the volume of the triangular prism below if $V = \frac{1}{2}bhl$?

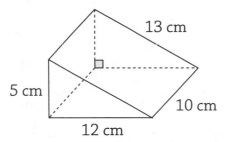

4. What is the volume of the cylinder if $V = \pi r^2 h$?

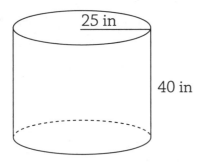

5. What is the volume of a sphere with a diameter of 7 cm if $V = \frac{4}{3}\pi r^3$?

6. What is the volume of a 5-inch-tall cone with a diameter of 3 inches
    if $V = \frac{1}{3}\pi r^2 h$?

7. What is the volume of the pyramid if $V = \frac{lwh}{3}$?

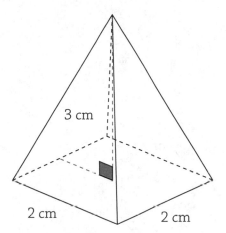

3 cm

2 cm   2 cm

EXERCISE 12-6

We've learned a good deal about manipulating figures. Let's put that new knowledge to use in answering the following questions.

1. Prasad has a digital photo that measures 4 inches by 6 inches. If he prints it out, it will not fit in his wallet, so he wants to reduce the size of the image. If he reduces the image by a scale factor of $\frac{1}{2}$, what will be the dimensions of the new photo?

2. Draw a new triangle on the coordinate plane with a scale factor of 1.5. Keep Point A at (1, 1).

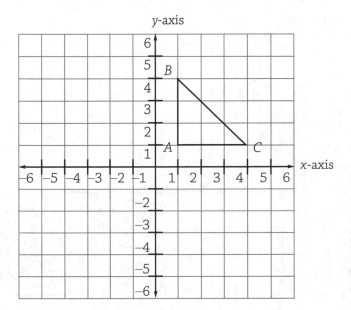

3. Circle the congruent figures.

4. If these two triangles are congruent, what is ∠EDF?

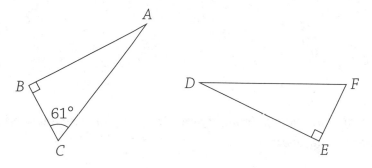

5. How many lines of symmetry does the figure have?

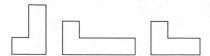

6. Draw a reflection of this triangle across the x-axis.

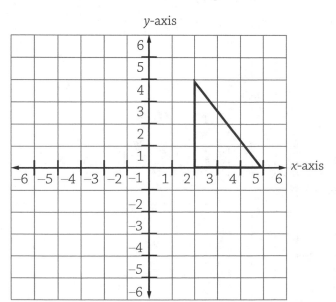

Answer Key

1
Number Properties

1. $41 + 98 = 139$
2. $156 + 75 = 231$
3. $9,872 + 12,340 = 22,212$
4. $39 - 23 = 16$
5. $482 - 264 = 218$
6. $7,947 - 3,608 = 4,339$

EXERCISE 1-2

1. $27 \times 46 = 1,242$
2. $352 \times 105 = 36,960$
3. $183 \div 4 = 45.75$
4. $405 \div 20 = 20.25$
5. $6,294 \div 48 = 131.125$

EXERCISE 1-3

1. $27 - 5 \times 3 + 1 = 27 - 15 + 1 = 12 + 1 = 13$
2. $(9 \div 3) + 4 \times 4 = 3 + 4 \times 4 = 3 + 16 = 19$

3. $6(4-2) - 3 \div 3 = 6(2) - 3 \div 3 = 12 - 3 \div 3 = 12 - 1 = 11$
4. $5 \times 3(3+2) = 5 \times 3(5) = 15(5) = 75$

EXERCISE 1-4

1. $|-46| = 46$
2. $|506| = 506$
3. $|-7| = 7$
4. $|-3.5| = 3.5$
5. $|0| = 0$

EXERCISE 1-5

1. $-14 + -13 = -27$
2. $23 + -72 = -49$
3. $-8 - (-3) = -5$
4. $6 - (-3) = 9$
5. $-9 + 0 = -9$, because any number plus 0 is that number.

EXERCISE 1-6

1. $7(-7) = -49$
2. $-4(-30) = 120$
3. $60 \div (-5) = -12$
4. $-18 \div (-3) = 6$
5. $-3 \times 0 \times -12 = 0$, because anything times 0 is 0.

Fractions

EXERCISE 2-1

1. $\dfrac{21}{15} = 1\dfrac{6}{15}$

2. $\dfrac{16}{8} = 2$

3. $\dfrac{25}{10} = 2\dfrac{5}{10} = 2\dfrac{1}{2}$

4. $\dfrac{17}{2} = 8\dfrac{1}{2}$

5. $\dfrac{34}{3} = 11\dfrac{1}{3}$

EXERCISE 2-2

1. $2\dfrac{1}{5} = \dfrac{11}{5}$

2. $1\dfrac{1}{8} = \dfrac{9}{8}$

3. $5\dfrac{2}{3} = \dfrac{17}{3}$

4. $17\dfrac{1}{2} = \dfrac{35}{2}$

5. $3\dfrac{3}{5} = \dfrac{18}{5}$

EXERCISE 2-3

1. $\dfrac{1}{5} = \dfrac{9}{45}$ and $\dfrac{2}{9} = \dfrac{10}{45}$, so $\dfrac{1}{5} < \dfrac{2}{9}$

2. $\dfrac{3}{8} = \dfrac{21}{56}$ and $\dfrac{4}{7} = \dfrac{32}{56}$, so $\dfrac{3}{8} < \dfrac{4}{7}$

3. $\dfrac{5}{15} = \dfrac{1}{3}$ and $\dfrac{4}{12} = \dfrac{1}{3}$, so $\dfrac{5}{15} = \dfrac{4}{12}$

4. $\dfrac{5}{6} = \dfrac{25}{30}$ and $\dfrac{4}{5} = \dfrac{24}{30}$, so $\dfrac{5}{6} > \dfrac{4}{5}$

5. $\dfrac{4}{16} = \dfrac{1}{4}$ and $\dfrac{3}{12} = \dfrac{1}{4}$, so $\dfrac{4}{16} = \dfrac{3}{12}$

EXERCISE 2-4

1. $\dfrac{1}{3} + \dfrac{1}{6} = \dfrac{2}{6} + \dfrac{1}{6} = \dfrac{3}{6} = \dfrac{1}{2}$

2. $\dfrac{2}{5} + \dfrac{1}{4} = \dfrac{8}{20} + \dfrac{5}{20} = \dfrac{13}{20}$

3. $\dfrac{2}{3} - \dfrac{1}{3} = \dfrac{1}{3}$

4. $\dfrac{3}{4} - \dfrac{2}{3} = \dfrac{9}{12} - \dfrac{8}{12} = \dfrac{1}{12}$

5. $\dfrac{4}{9} - \dfrac{1}{4} = \dfrac{16}{36} - \dfrac{9}{36} = \dfrac{7}{36}$

EXERCISE 2-5

1. $\dfrac{1}{2} \times \dfrac{3}{4} = \dfrac{3}{8}$

2. $\dfrac{5}{6} \times \dfrac{4}{7} = \dfrac{20}{42} = \dfrac{10}{21}$

3. $\dfrac{1}{4} \div \dfrac{7}{8} = \dfrac{1}{4} \times \dfrac{8}{7} = \dfrac{8}{28} = \dfrac{2}{7}$

4. $\dfrac{5}{8} \div \dfrac{1}{4} = \dfrac{5}{8} \times \dfrac{4}{1} = \dfrac{20}{8} = 2\dfrac{4}{8} = 2\dfrac{1}{2}$

5. $\dfrac{3}{7} \div \dfrac{3}{1} = \dfrac{3}{7} \times \dfrac{1}{3} = \dfrac{3}{21} = \dfrac{1}{7}$

Decimals

EXERCISE 3-1

1. In 3.468, the 3 is in the ones place.
2. In 0.317, the 3 is in the tenths place.
3. In 1.8673, the 3 is in the ten-thousandths place.
4. In 13,067.84, the 3 is in the thousands place.
5. In 2.63, the 3 is in the hundredths place.

EXERCISE 3-2

1. $72.1 > 7.027 > 7.021$
2. $0.967 > 0.9067 > 0.0967$
3. $1.293 > 1.2906 > 1.2904$
4. $0.0846 > 0.084 > 0.08$
5. $263.6 > 263.16 > 2.636$

EXERCISE 3-3

1. $34.56 + 23.98 = 58.54$
2. $575.234 + 1{,}290.35 = 1{,}865.584 = 1{,}865.58$
3. $2.3 - 2.1 = 0.2 = 0.20$
4. $0.93485 - 0.02348 = 0.91137 = 0.91$
5. $6.23 - 0.023 = 6.207 = 6.21$

EXERCISE 3-4

1. $23.3 \times 0.02 = 0.466$
2. $568 \times 0.1 = 56.8$
3. $13.5 \times 1.9 = 25.65$
4. $654.89 \times 1.7 = 1{,}113.313$
5. $84.6 \times 100 = 8{,}460$

EXERCISE 3-5

1. $52.6 \div 4 = 13.15 = 13.2$
2. $8 \div 4 = 20 = 20.0$. Using a decimal point and another 0 shows that this is a precise measurement to three significant digits.
3. $23.6 \div 1.6 = 14.75 = 14.8$
4. $98.7 \div 0.1 = 987$
5. $762 \div 100 = 7.62$

EXERCISE 3-6

1. $\dfrac{1}{8} = 8\overline{)1.000}$

$$
\begin{array}{r}
.125 \\
8\,\overline{)1.000} \\
\underline{8} \\
20 \\
\underline{16} \\
40 \\
\underline{40} \\
0
\end{array}
$$

$\dfrac{1}{8} = 0.125$. Round to tenths place: 0.1.

2. $\dfrac{17}{18} = 18\overline{)17.000}$

$$
\begin{array}{r}
.944 \\
18\,\overline{)17.000} \\
\underline{162} \\
80 \\
\underline{72} \\
80
\end{array}
$$

> You can see this will keep repeating, so round off or show it as a repeating decimal.

$\dfrac{17}{18} = 0.94$, or $0.9\overline{4}$. Round to tenths place: 0.9.

3. $\dfrac{6}{10}$

Tenths is a power of 10, so remember that when you divide by 10, you can just move the decimal point one place to the left.

$6.0 \rightarrow 0.6$

$\dfrac{6}{10} = 0.6$

4.

$$
\begin{array}{r}
.3125 \\
16\overline{)5.0000} \\
\underline{48} \\
20 \\
\underline{16} \\
40 \\
\underline{32} \\
80 \\
\underline{80} \\
0
\end{array}
$$

$\dfrac{5}{16} = 0.3125$. Round to tenths place: 0.3.

5. $\dfrac{34}{100}$

Since 100 is power of 10, you can move the decimal 2 places to the left: $34 \rightarrow 0.34$. Round to tenths place: 0.3.

EXERCISE 3-7

1. $0.8 = \dfrac{8}{10}$

2. $1.2 = \dfrac{12}{10} = \dfrac{6}{5}$

3. $345.23 = \dfrac{34{,}523}{100}$

4. $8.2345 = \dfrac{82{,}345}{10{,}000} = \dfrac{16{,}469}{2{,}000}$

5. $105.00 = \dfrac{10,500}{100} = \dfrac{105}{1}$

Percents

EXERCISE 4-1

To convert percents to decimals, move the decimal point 2 places to the left.

1. $30\% = \dfrac{30}{100} = 0.3$

2. $16\% = \dfrac{16}{100} = 0.16$

3. $2.5\% = \dfrac{2.5}{100} = 0.025$

4. $120\% = \dfrac{120}{100} = 1.2$

5. $0.5\% = \dfrac{0.5}{100} = 0.005$

EXERCISE 4-2

To convert decimals to percents, move the decimal point 2 places to the right.

1. $0.38 = 38\%$

2. $0.539 = 53.9\%$

3. $1.246 = 124.6\%$

4. $0.06 = 6\%$

EXERCISE 4-3

To convert a percent to a fraction, put the percent over 100 and reduce if needed.

1. $30\% = \dfrac{30}{100} = \dfrac{3}{10}$

2. $2.5\% = \dfrac{2.5}{100} = \dfrac{25}{1{,}000} = \dfrac{1}{40}$

3. $120\% = \dfrac{120}{100} = \dfrac{12}{10} = \dfrac{6}{5}$, or $1\dfrac{1}{5}$

4. $0.5\% = \dfrac{0.5}{100} = \dfrac{5}{1{,}000} = \dfrac{1}{200}$

EXERCISE 4-4

To convert a fraction to a percent, increase the fraction to have 100 in the denominator or convert the fraction to a decimal and then to a percent.

1. $\dfrac{1}{8} = 0.125 = 12.5\%$

2. $\dfrac{1}{3} = 0.\overline{3} = 33\overline{3}\%$

3. $\dfrac{3}{20} \times \dfrac{5}{5} = \dfrac{15}{100} = 15\%$

4. $\dfrac{2}{7} = 0.286 = 29\%$

5. $\dfrac{2}{5} \times \dfrac{20}{20} = \dfrac{40}{100} = 40\%$

EXERCISE 4-5

1. $\dfrac{30}{100} \times \dfrac{520}{1} = 156$

   or $0.3 \times 520 = 156$

2. $\dfrac{28}{100} \times \dfrac{10}{1} = \dfrac{280}{100} = 2.8$

   or $0.28 \times 10 = 2.8$

3. $\dfrac{10}{100} \times \dfrac{28}{1} = \dfrac{280}{100} = 2.8$

   or $10 \times 0.28 = 2.8$

4. $\dfrac{6.5}{100} \times \dfrac{2}{1} = \dfrac{13}{100} = 0.13$

5. $\dfrac{70}{100} \times \dfrac{20}{1} = \dfrac{1,400}{100} = 14$ invited

   or $0.7 \times 20 = 14$ invited $\rightarrow 20 - 14 = 6$ *not* invited

EXERCISE 4-6

To find the amount of interest, multiply the interest rate times the principal times the time in years: $I = p \times r \times t$.

1. $\dfrac{200}{1} \times \dfrac{4}{100} \times 1 = \dfrac{800}{100} = 8$

   or $200 \times 0.04 \times 1 = 8$

2. 3 months is $\dfrac{1}{4}$ of a year, so $t = \dfrac{1}{4}$.

   $\dfrac{400}{1} \times \dfrac{3}{100} \times \dfrac{1}{4} = \dfrac{1,200}{400} = 3$

   or $400 \times 0.03 \times 0.25 = 3$

3. $\dfrac{1,000}{1} \times \dfrac{14}{100} \times 1 = \dfrac{14,000}{100} = 140$

   or $1,000 \times 0.14 \times 1 = 140$

4. Find the interest: $\dfrac{600}{1} \times \dfrac{9}{100} \times 2 = \dfrac{10,800}{100} = 108$

   or $600 \times 0.09 \times 2 = 108$

   Add the interest to the principal: $\$600 + 108 = \$708$.

EXERCISE 4-7

1. $\dfrac{25}{100} \times \dfrac{60}{1} = \dfrac{1}{4} \times \dfrac{60}{1} = \dfrac{60}{4} = \$15$ or $0.25 \times \$60 = \$15$

   Subtract to find the new price: $\$60 - \$15 = \$45$. We could also just find 75% of 60 to get the new price in one step:

   $\dfrac{75}{100} \times \dfrac{60}{1} = \dfrac{3}{4} \times \dfrac{60}{1} = \dfrac{180}{4} = \$45$ or $0.75 \times \$60 = \$45$

2. $42 \times 0.3 = \$12.60$ or $\dfrac{30}{100} \times \dfrac{42}{1} = \dfrac{1,260}{100} = \$12.60$

   Subtract to find the new price: $\$42 - \$12.60 = \$29.40$. We could also just find 70% of 42 to get the new price in one step. Doing 70% as a decimal will probably be easier than using a fraction: $42 \times 0.7 = \$29.40$.

3. $\dfrac{12}{100} \times \dfrac{90,000}{1} = \$10,800$ or $0.12 \times 90,000 = \$10,800$

4. $\dfrac{8}{100} \times \dfrac{12}{1} = \dfrac{96}{100} = 0.96$ or $0.08 \times 12 = \$0.96$

5. To get a 15%, let's find 10% and 5%. 10% of 21 is 2.1. Half of that is 1.05. Add those together to get $3.15. Add that tip to the $21 to get $24.15.

6. Find 10% and double it. 10% of 40 is 4. Double that to get $8. Add that tip to the $40 to get $48.00.

EXERCISE 4-8

Use the percent change formula $\dfrac{\text{difference}}{\text{original}}$ and then convert the decimal to a percent.

1. The difference is $130{,}000 - 100{,}000 = 30{,}000$.

   $$\frac{30{,}000}{100{,}000} = \frac{3}{10} = 0.3 = 30\%$$

2. The difference is $14 - 12 = 2$.

   $$\frac{2}{12} = 0.1\overline{6} = 17\%$$

3. The difference is $6 - 5 = 1$. The question says, "percent greater," so the original number must be the smaller one, 5.

   $$\frac{1}{5} = 0.2 = 20\%$$

4. The difference is $50{,}000 - 1{,}400 = 48{,}600$.

   $$\frac{48{,}600}{50{,}000} = \frac{486}{500} = 0.972 = 97\%$$

Ratios and Proportions

EXERCISE 5-1

1. Write the ratio given and reduce it as you would with a fraction. If you have trouble finding the largest common multiple to divide the fraction by, just start by dividing by 2 and then keep going:

   $$\frac{400\,\text{students}}{32\,\text{teachers}} = \frac{200}{16} = \frac{100}{8} = \frac{50}{4} = \frac{25}{2}$$

The ratio of students to teachers is 25:2 or we can say 12.5:1.

2. Since the question asks for the ratio of hamburgers to hot dogs, be sure to put hamburgers on top!

$$\frac{80 \text{ hamburgers}}{100 \text{ hot dogs}} = \frac{80}{100} = \frac{8}{10} = \frac{4}{5}$$

The ratio of hamburgers to hot dogs is 4:5.

3. To find the ratio of tan kittens to all kittens, we need the total: 2 white + 3 tan = 5 total, so the ratio of tan kittens to all kittens is 3:5.

4. We might want to set up a chart for this one since we will need to find the multiplier.

	Ratio	Multiplier	Actual
students	20		640
teachers	1		

To find the multiplier, divide 640 by 20:

$$\frac{640}{20} = \frac{64}{2} = 32$$

Multiply that by the ratio number for teachers.

	Ratio	Multiplier	Actual
students	20	32	640
teachers	1	32	32

There should be 32 teachers.

5. Set up a chart. Make sure to draw a column for total since that is involved.

	Ratio	Multiplier	Actual
chocolate chip	12		
sugar	8		
total			100

If there are 12 parts chocolate chip cookies to 8 parts sugar cookies, the total number of parts is 20. Fill that into the chart and find the multiplier. Divide 100 by 20 to get 5. Write that in as the multiplier and solve for the other actual amounts.

	Ratio	Multiplier	Actual
chocolate chip	12	5	60
sugar	8	5	40
total	20	5	100

There are 40 sugar cookies. There are 60 chocolate chip cookies.

6. Make a chart. Use three columns for the three ingredients. Since the total is not involved, you can leave that column out if you like.

	Ratio	Multiplier	Actual
flour	3		
butter	2		1 cup
sugar	1		

Divide 1 by 2 to get the multiplier: $\frac{1}{2}$. Multiply the amounts of flour and sugar by $\frac{1}{2}$ to find the actual amounts.

	Ratio	Multiplier	Actual
flour	3	$\frac{1}{2}$	$1\frac{1}{2}$ cups
butter	2	$\frac{1}{2}$	1 cup
sugar	1	$\frac{1}{2}$	$\frac{1}{2}$ cup

EXERCISE 5-2

1. $\dfrac{5}{25} = \dfrac{2}{x}$

Cross multiply: $5x = 2 \times 25$, so $5x = 50$. Divide both sides by 5 to solve for x:

$\dfrac{5x}{5} = \dfrac{50}{5}$, so $x = 10$

2. $\dfrac{x}{3} = \dfrac{9}{54}$

Cross multiply: $54x = 9 \times 3$, so $54x = 27$. Divide both sides by 54 to solve for x:

$\dfrac{54x}{54} = \dfrac{27}{54}$, so $x = \dfrac{27}{54} = \dfrac{1}{2}$

3. $\dfrac{20 \text{ teachers}}{100 \text{ students}} = \dfrac{x \text{ teachers}}{30 \text{ students}}$

Cross multiply: $100x = 20 \times 30$, so $100x = 600$. Divide both sides by 100 to solve for x:

$\dfrac{100x}{100} = \dfrac{600}{100}$, so $x = \dfrac{600}{100} = 6$

4. $\dfrac{104\,\text{apples}}{x\,\text{oranges}} = \dfrac{13\,\text{apples}}{8\,\text{oranges}}$

Cross multiply: $13x = 104 \times 8$ so $13x = 832$. Divide both sides by 13 to solve for x:

$\dfrac{13x}{13} = \dfrac{832}{13}$, so $x = \dfrac{832}{13} = 64$

5. Set up a proportion. Let's use x for the missing value:

$\dfrac{2\,\text{eggs}}{24\,\text{cupcakes}} = \dfrac{x}{36\,\text{cupcakes}}$

Cross multiply: $24x = 2 \times 36$, so $24x = 72$. Divide both sides by 24 to solve for x:

$\dfrac{24x}{24} = \dfrac{72}{24}$, so $x = \dfrac{72}{24} = 3$

Landon will need 3 eggs to make 36 cupcakes.

6. Set up a proportion. Let's use x for the missing value:

$\dfrac{2\,\text{hours}}{68\,\text{bricks}} = \dfrac{3\,\text{hours}}{x\,\text{bricks}}$

Cross multiply: $2x = 68 \times 3$, so $2x = 204$. Divide both sides by 2 to solve for x:

$\dfrac{2x}{2} = \dfrac{204}{2}$, so $x = \dfrac{204}{2} = 102$

Molly can lay 102 bricks in 3 hours.

7. Set up a proportion. Let's use x for the missing value:

$\dfrac{6\,\text{inches}}{15.24\,\text{cm}} = \dfrac{8\,\text{inches}}{x\,\text{cm}}$

Cross multiply: $6x = 15.24 \times 8$, so $6x = 121.92$. Divide both sides by 6 to solve for x:

$$\frac{6x}{6} = \frac{121.92}{6}, \text{ so } x = \frac{121.92}{6} = 20.32$$

8 inches is 20.32 centimeters.

8. Set up a proportion. Let's use x for the missing value:

$$\frac{4 \text{ inches}}{6 \text{ inches}} = \frac{x \text{ inches}}{8 \text{ inches}}$$

Cross multiply: $6x = 4 \times 8$, so $6x = 32$. Divide both sides by 6 to solve for x.

$$\frac{6x}{6} = \frac{32}{6}, \text{ so } x = \frac{32}{6} = 5\frac{1}{3}$$

The width will be $5\frac{1}{3}$ inches.

EXERCISE 5-3

1. Set up a proportion. Let's use x for the missing value:

$$\frac{11 \text{ hot dogs}}{5 \text{ minutes}} = \frac{20 \text{ hot dogs}}{x \text{ minutes}}$$

Cross multiply: $11x = 5 \times 20$, so $11x = 100$. Divide both sides by 11 to solve for x:

$$\frac{11x}{11} = \frac{100}{11}, \text{ so } x = \frac{100}{11} = 9.\overline{09}, \text{ but we can round that to 9.1.}$$

It will take Bruce 9.1 minutes to eat 20 hot dogs.

2. Set up a proportion. Let's use x for the missing value:

$$\frac{1{,}100 \text{ stitches}}{1 \text{ minute}} = \frac{x \text{ stitches}}{7 \text{ minutes}}$$

Cross multiply: $x = 1{,}100 \times 7$, so $x = 7{,}700$. The machine can sew 7,700 stitches in 7 minutes.

3. Set up a proportion. Let's use x for the missing value:

$$\frac{\$13.40}{1 \text{ hour}} = \frac{x}{6 \text{ hours}}$$

Cross multiply: $x = 13.40 \times 6$, so $x = 80.4$. Vicki earns $80.40 on Tuesday.

4. Set up a proportion. Let's use x for the missing value:

$$\frac{78 \text{ loaves}}{2 \text{ hours}} = \frac{x \text{ loaves}}{1 \text{ hour}}$$

Cross multiply: $2x = 78 \times 1$, so $2x = 78$ and $x = 39$. WINK, Inc. makes 39 loaves per hour.

5. We need to find the unit rate for each store, the price per apple.

Store A: $\dfrac{\$1}{12 \text{ apples}} = \dfrac{x}{1 \text{ apple}}$

Cross multiply: $12x = 4 \times 1$, so $12x = 4$ and $x = \$0.33$

Store B: $\dfrac{\$9}{20 \text{ apples}} = \dfrac{x}{1 \text{ apple}}$

Cross multiply: $20x = 9 \times 1$, so $20x = 9$ and $x = \$0.45$. One apple costs $0.33 at Store A and $0.45 at Store B. Store A has the better deal.

6. We need to find the unit rate for each runner, the number of minutes per kilometer.

Ryan: $\dfrac{5\,\text{km}}{23\,\text{minutes}} = \dfrac{1\,\text{km}}{x\,\text{minutes}}$

Cross multiply: $5x = 23$ and $x = 4.6$.

Chris: $\dfrac{3\,\text{km}}{17\,\text{minutes}} = \dfrac{1\,\text{km}}{x\,\text{minutes}}$

Cross multiply: $3x = 17$ and $x = 5.7$. Ryan is faster. His rate is 4.6 minutes per kilometer.

The Coordinate Plane

EXERCISE 6-1

1 – 6.

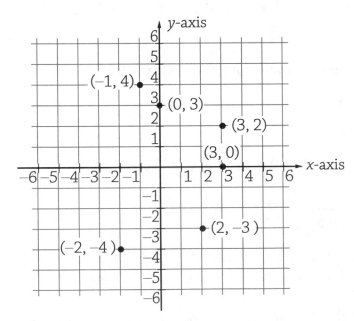

7. Point A is at $(3, -4)$ in Quadrant IV.

8. Point B is at $(-2, 1)$ in Quadrant II.

EXERCISE 6-2

1.

x	y
−3	−2
−2	−1
−1	0
0	1
1	2

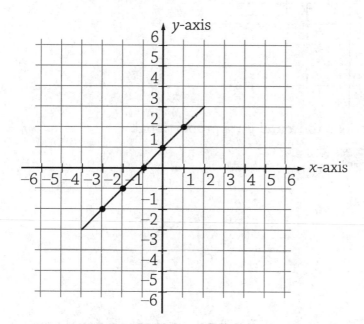

This is a linear function, but *x* and *y* are not proportional.

2.

x	y
−6	−2
−3	−1
0	0
3	1
6	2

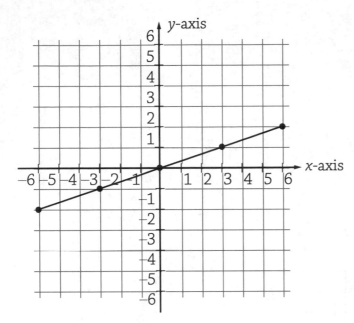

This is a linear function, and **x** and **y** are proportional.

3.

x	y
2	−2
1	−1
0	0
−1	1
−2	2

The equation of the line is $x = -y$. The constant of proportionality is -1.

EXERCISE 6-3

1. The car has gone 100 miles after $2\frac{1}{2}$ hours.

2. It takes the car 1 hour to go 40 miles.

3. Lynda has read 20 pages in 1 hour. Remember to convert the minutes shown on the chart into hours.

4. It takes Alex 90 minutes to read 45 pages.

5. Alex is reading faster.

EXERCISE 6-4

1. Find the point at which $y = 1$. The x-coordinate of that point is the unit rate. For this graph, when y equals 1, x also equals 1, so the unit rate is 1 cookie per minute. Yum!

2. The graph for 0.5 mile per hour should look like this:

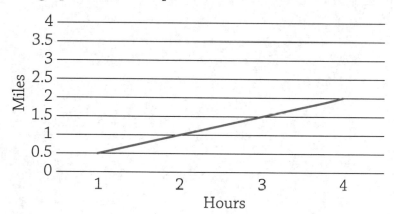

Exponents and Roots

EXERCISE 7-1

1. $3 \times 3 \times 3 = 3^3$
2. $5 \times 5 \times 5 \times 5 \times 5 \times 5 = 5^6$
3. $234 \times 234 = 234^2$
4. $78^4 = 78 \times 78 \times 78 \times 78$
5. $2^6 = 2 \times 2 \times 2 \times 2 \times 2 \times 2$
6. $6^3 = 6 \times 6 \times 6$

EXERCISE 7-2

1. $2^3 \times 2^5 = 2^8$
2. $54^2 \times 54^9 = 54^{11}$
3. $6 \times 6^{15} = 6^{16}$
4. $4^{-3} \times 4^3 = 4^0 = 1$

EXERCISE 7-3

1. $\dfrac{7^4}{7^2} = 7^2$

2. $\dfrac{9^8}{9^8} = 9^0 = 1$

3. $\dfrac{27^{13}}{27^5} = 27^8$

4. $\dfrac{12^3}{12^6} = 12^{-3} = \dfrac{1}{12^3}$

EXERCISE 7-4

1. $(5^3)^4 = 5^{12}$

2. $(11^8)^1 = 11^8$

3. $(y^2)^5 = y^{10}$

4. $(5x^3 y)^4 = 5^4 x^{12} y^4$

EXERCISE 7-5

1. $\dfrac{1}{25} = \dfrac{1}{5^2} = 5^{-2}$

2. $\dfrac{1}{81} = \dfrac{1}{9^2} = 9^{-2}$

3. $\dfrac{1}{8} = \dfrac{1}{2^3} = 2^{-3}$

4. $3^{-2} = \dfrac{1}{3^2} = \dfrac{1}{9}$

5. $41^{-7} = \dfrac{1}{41^7}$

6. $1^{-5} = \dfrac{1}{1^5} = \dfrac{1}{1} = 1$

EXERCISE 7-6

1. $\sqrt{36} = 6$
2. $\sqrt{100} = 10$
3. $\sqrt{49} = 7$
4. $\sqrt{54} = \sqrt{9 \times 6} = 3\sqrt{6}$
5. $\sqrt{44} = \sqrt{4 \times 11} = 2\sqrt{11}$
6. $\sqrt{98} = \sqrt{2 \times 49} = 7\sqrt{2}$
7. $\sqrt{50} \approx 7.1$, since $\sqrt{49} = 7$
8. $\sqrt{110} \approx 10.5$, since 110 is about halfway between 100 and 121 and $\sqrt{100} = 10$ and $\sqrt{121} = 11$
9. $\sqrt{80} \approx 8.9$, since $\sqrt{81} = 9$
10. $\sqrt{810} \approx 28$, since $\sqrt{810} = \sqrt{81 \times 10} = 9\sqrt{10}$ and $\sqrt{10} \approx 3.1$, so $9 \times 3.1 = 27.9$

EXERCISE 7-7

1. $4\sqrt{5} + 2\sqrt{5} = 6\sqrt{5}$
2. $\sqrt{5} + 12\sqrt{5} = 13\sqrt{5}$

3. $6\sqrt{3} - 3\sqrt{3} = 3\sqrt{3}$

4. $5\sqrt{2} - \sqrt{2} = 4\sqrt{2}$

EXERCISE 7-8

1. $\sqrt{2} \times \sqrt{3} = \sqrt{6}$

2. $6\sqrt{2} \times 2\sqrt{5} = 12\sqrt{10}$

3. $\sqrt{5} \times \sqrt{5} = \sqrt{25} = 5$

4. $2\sqrt{8} \times \sqrt{20} = 2\sqrt{160} = 2\sqrt{16 \times 10} = 2 \times 4\sqrt{10} = 8\sqrt{10}$

EXERCISE 7-9

1. $\sqrt{6} \div \sqrt{3} = \sqrt{\dfrac{6}{3}} = \sqrt{2}$

2. $\sqrt{8} \div \sqrt{2} = \sqrt{\dfrac{8}{2}} = \sqrt{4} = 2$

3. $\left(4\sqrt{5}\right) \div \left(2\sqrt{5}\right) = \dfrac{4\sqrt{5}}{2\sqrt{5}} = \dfrac{4}{2}\sqrt{\dfrac{5}{5}} = 2\sqrt{1} = 2$

4. $\left(12\sqrt{5}\right) \div \left(7\sqrt{2}\right) = \dfrac{12\sqrt{5}}{7\sqrt{2}} = \dfrac{12}{7}\sqrt{\dfrac{5}{2}} = \dfrac{12\sqrt{5}}{7\sqrt{2}}$. We cannot leave a root in

   the denominator of a fraction, so multiply the fraction by 1 in the form

   of $\dfrac{\sqrt{2}}{\sqrt{2}}$: $\dfrac{12\sqrt{5}}{7\sqrt{2}} \times \dfrac{\sqrt{2}}{\sqrt{2}} = \dfrac{12\sqrt{10}}{7\sqrt{4}} = \dfrac{12\sqrt{10}}{7 \times 2} = \dfrac{12\sqrt{10}}{14}$. We can simplify

   that to $\dfrac{6\sqrt{10}}{7}$.

EXERCISE 7-10

1. $94{,}000{,}000 = 9.4 \times 10^{7}$

2. $2{,}600 = 2.6 \times 10^3$

3. $0.83 = 8.3 \times 10^{-1}$

4. $0.0000524 = 5.24 \times 10^{-5}$

5. $4.83 \times 10^7 = 48{,}300{,}000$

6. $9.6 \times 10^2 = 960$

7. $5.462 \times 10^{-4} = 0.0005462$

8. $8.4 \times 10^{-1} = 0.84$

Equations and Inequalities

EXERCISE 8-1

1. $x - 11$

2. $2(ab)$

3. $\dfrac{x}{7} + 5$

EXERCISE 8-2

1. the sum of a number and 106

2. twelve divided by a number

3. three and a half subtracted from the product of 4 and a number

EXERCISE 8-3

1.
$$x - 8 = 12$$
$$x - 8 + 8 = 12 + 8$$
$$x = 20$$

2.
$$5 + x = 7$$
$$5 + x - 5 = 7 - 5$$
$$x = 2$$

3.
$$\frac{x}{3} = 4$$
$$\frac{x}{3} \times 3 = 4 \times 3$$
$$x = 12$$

4. $7x = 49$
$$\frac{7x}{7} = \frac{49}{7}$$
$$x = 7$$

5.
$$\frac{10}{x} = 2$$
$$\frac{10}{x} \times \frac{x}{1} = 2 \times \frac{x}{1}$$
$$10 = 2x$$
$$\frac{10}{2} = \frac{2x}{2}$$
$$5 = x, \text{ so } x = 5$$

6.
$$4x + 6 = 38$$
$$4x + 6 - 6 = 38 - 6$$
$$4x = 32$$
$$\frac{4x}{4} = \frac{32}{4}$$
$$x = 8$$

7.
$$3(x + 2) = 3x + 11$$
$$3x + 6 = 3x + 11$$
$$3x + 6 - 3x = 3x + 11 - 3x$$
$$6 \neq 11, \text{ so this equation has no solution}$$

8. $\dfrac{2}{x} + 5 = 6$

$\dfrac{2}{x} + 5 - 5 = 6 - 5$

$\dfrac{2}{x} = 1$

$\dfrac{2}{x} \times \dfrac{x}{1} = 1 \times \dfrac{x}{1}$

$2 = 1x$, so $x = 2$

9. $x + y = 14$

$x + y - y = 14 - y$

$x = 14 - y$

10. $4(2 + x) = 4x + 8$

$8 + 4x = 4x + 8$

$8 + 4x - 8 = 4x + 8 - 8$

$4x = 4x$, so this equation has infinite solutions.

EXERCISE 8-4

1. $x + 4 < 7$

$x + 4 - 4 < 7 - 4$

$x < 3$

2. $-2 + x > 9$

$-2 + x + 2 > 9 + 2$

$x > 11$

3. $\dfrac{x}{5} \le 4$

$\dfrac{x}{5} \times 5 \le 4 \times 5$

$x \le 20$

4. $12x \geq 40$

$$\frac{12x}{12} \geq \frac{40}{12}$$

$$x \geq \frac{40}{12}$$

$$x \geq \frac{10}{3} \text{ or } 3\frac{1}{3}$$

5. $\dfrac{8}{x} < 1$

$$\frac{8}{x} \times \frac{x}{1} < 1 \times \frac{x}{1}$$

$$8 < x, \text{ so } x > 8$$

6. $3x - 4 > 86$

$$3x - 4 + 4 > 86 + 4$$

$$3x > 90$$

$$\frac{3x}{3} > \frac{90}{3}$$

$$x > 30$$

7. $\dfrac{20}{x} + 3 \geq 7$

$$\frac{20}{x} + 3 - 3 \geq 7 - 3$$

$$\frac{20}{x} \geq 4$$

$$\frac{20}{x} \times \frac{x}{1} \geq 4 \times \frac{x}{1}$$

$$20 \geq 4x$$

$$\frac{20}{4} \geq \frac{4x}{4}$$

$$5 \geq x, \text{ so } x \leq 5$$

8.
$$-2x - 4 \le 3x + 6$$
$$-2x - 4 + 4 \le 3x + 6 + 4$$
$$-2x \le 3x + 10$$
$$-2x - 3x \le 3x + 10 - 3x$$
$$-5x \le 10$$
$$\frac{-5x}{-5} \le \frac{10}{-5}$$

Since we divided by a negative, we need to flip the sign: $x \ge -2$.

EXERCISE 8-5

1. $x = 5$ and $y = 6$. Solve the first equation for x in terms of y: $x = 11 - y$. Substitute that into the second equation:

$$3(11 - y) - y = 9$$
$$33 - 4y = 9$$
$$-4y = -24$$
$$y = 6$$

Plug that into the first equation to solve for x: $x + 6 = 11$, so $x = 5$.

2. $x = 5$ and $y = 6$. Add the two equations:

$$x + y = 11$$
$$+(3x - y = 9)$$
$$4x = 20$$
$$x = 5$$

Plug that into the first equation to solve for y: $5 + y = 11$, so $y = 6$.

3. $x = 5$ and $y = 6$.

For $x + y = 11$:

x	y
1	10
3	8
5	6

For $3x - y = 9$:

x	y
1	−6
3	0
5	6

Graph both lines:

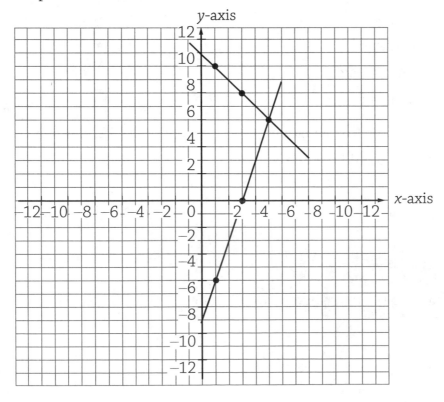

The intersection of the lines on the graph is at (5, 6), which means $x = 5$ and $y = 6$.

Data Presentation

EXERCISE 9-1

1. Raegan. Raegan sold 6 boxes on Day 5, and Hayleigh sold 5 boxes.
2. Day 2. Each person sold 5 boxes.
3. Hayleigh. Add up the total for each person. Hayleigh sold 34 boxes, and Raegan sold 26.

EXERCISE 9-2

1. Ali. Ali read 5 books, while Jay read 2 and Bella read 4.
2. June. Jay and Ali each read 2 books in June.
3. Ali. Ali read 2 books in June, 4 books in July, and 5 books in August.

EXERCISE 9-3

1. Entertainment. The smallest amount is 3%, which was spent on entertainment.
2. Food and phone. Utilities is 12%. Phone is 8% and food is 4%, so those two categories together are 12%.
3. $500: $\frac{25}{100} \times 2,000 = 25 \times 20 = 500$

EXERCISE 9-4

1.

Stem	Leaf
3	6
4	2, 5, 7, 8, 9
5	0, 5, 5, 9
6	1, 2, 3, 8

2. 32. The highest score was 67, and the lowest score was 35. $67 - 35 = 32$.

3. 53. There are 13 scores, so the 7th score will be the median.

4. 2 games. There are two scores above 60: 62 and 67.

EXERCISE 9-5

1. 31. The line marked inside the box represents the median age, which is 31.

2. 10 years old. The bottom line represents the smallest data point, which is 10.

3. 12. The third quartile begins at the median, 31, and goes up to the top of the box at 43. $43 - 31 = 12$.

EXERCISE 9-6

1. 1 person. Only one person is shown above the 46 mark, in the range of 67 to 71 sit-ups.

2. 2 people. There is one person in the 27–31 range and one in the 32–36 range.

3. 12 people. Add up the number of people in each category:
$3 + 2 + 1 + 1 + 2 + 2 + 0 + 0 + 0 + 0 + 1 = 12$.

EXERCISE 9-7

1. 9. Count the ends of the branches. There are nine possible outcomes.

2. 3. There are three outcomes in which the same color marble is chosen both times: red/red, white/white, and black/black.

3.

thick crust	pepperoni
	sausage
	cheese
thin crust	pepperoni
	sausage
	cheese

EXERCISE 9-8

1. 25. A total of 36 people can ice-skate, but 11 of them also roller-skate, so we need to subtract those people: $36 - 11 = 25$.

2. 74. A total of 49 people can roller-skate, and a total of 36 people can ice-skate: $49 + 36 = 85$. However, the people who can do both have been counted twice (once with roller-skaters and once with ice-skaters) so we need to subtract 11. $85 - 11 = 74$.

3.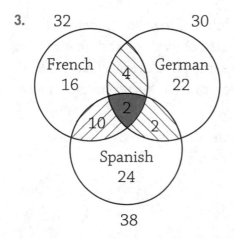

32   30

French   4   German
16   22
    10   2   2
    Spanish
      24

38

EXERCISE 9-9

1. Graph *A*. Graph *A* shows a dot at 0 for an outlier. Graph *B* shows 1 student in the 0–10 range, but there is no way to know the actual number of pages that the student read.

2. Both graphs clearly show a gap in the data in the 10- to 19-page range.

3. Graph *A*, 35. Graph *A* makes the median easier to find because it has a line at that spot. The median is 35.

4. Graph *B*, 6. Graph *B* makes it easy to see how many students read a certain range of pages. Graph *A* does not show the numbers of students for any range. The 30- to 39-page range has 6 students.

5. Graph *B*, 16. Only Graph *B* has the numbers of students listed. Add up the number of students for each bar: $1 + 0 + 5 + 6 + 4 = 16$.

Statistics

EXERCISE 10-1

1. *C*. Choices *B*, *C*, and *D* all survey the residents of Iowa, so they are all better choices than *A*. Choice *D* runs the risk of over- or underestimating if the counties don't all have roughly the same populations. Choices *B* and *C* both provide representative samples, but *C* is better because more people are surveyed.

2. Selection bias. By surveying people who are already *in* a candy store, Aaron is very likely to find that they like candy.

3. Self-selection bias and observer bias. The people surveyed are the ones listening to the podcast *Trivial*, so they are more likely to enjoy it. Since the survey is done by the host of *Trivial*, people may be influenced to say what they think the host wants to hear.

EXERCISE 10-2

1. 18.25. Add up the eight values and divide the total by 8:
   $15 + 9 + 23 + 8 + 42 + 6 + 35 + 8 = 146$ and $146 \div 8 = 18.25$.

2. 12. Put the data in order: {6, 8, 8, 9, 15, 23, 35, 42}. Since there are two middle values, we average them to find the median. $9 + 15 = 24$ and $24 \div 2 = 12$.

3. 8. The value 8 appears twice, and that is more times than any other value, so 8 is the mode.

4. 36. Subtract the smallest value from the largest: $42 - 6 = 36$. The range is 36.

5. 25. Currently, the average is 18.25, which is 146 points divided by 8 values. Let's think about this as an equation. Average = total of values divided by number of values: $A = t \div n$. What we have now is $18.25 = 146 \div 8$. What we want is $19 = t \div 9$. It's $t \div 9$ because if we add a 9th value, we will be dividing the total number of points by 9 instead of by 8. Solve the equation for t:

$$19 = \frac{t}{9}$$

Multiply both sides by 9 to isolate t.

$$(9)19 = \frac{t}{9} \times 9$$

$$171 = t$$

We need 171 total points. We have 146, so we need $171 - 146 = 25$ more points to raise the average to 19.

EXERCISE 10-3

1. 88.8. Add up Bradley's five scores and divide the total by 5:
   $85 + 87 + 91 + 89 + 92 = 444$ and $444 \div 5 = 88.8$.

2. 84. Add up Audree's five scores and divide the total by 5:
   $96 + 73 + 86 + 90 + 75 = 420$ and $420 \div 5 = 84$.

3. 2.24. Find the positive difference between each of Bradley's scores and
   the mean: $88.8 - 85 = 3.8$; $88.8 - 87 = 1.8$; $91 - 88.8 = 2.2$;
   $89 - 88.8 = 0.2$; $92 - 88.8 = 3.2$.

   Then average those differences: $3.8 + 1.8 + 2.2 + 0.2 + 3.2 = 11.2$
   and $11.2 \div 5 = 2.24$

4. 8. Find the positive difference between each of Audree's scores and the
   mean: $96 - 84 = 12$; $84 - 73 = 11$; $86 - 84 = 2$; $90 - 84 = 6$;
   $84 - 75 = 9$.

   Then average those differences: $12 + 11 + 2 + 6 + 9 = 40$
   and $40 \div 5 = 8$.

5. Bradley has the more consistent performance because the mean average
   deviation of his scores is smaller.

EXERCISE 10-4

1. 25%. A deck of 52 cards has 4 suits of 13 cards each, so there are 13
   hearts out of 52 cards. $\dfrac{13}{52} = \dfrac{1}{4} = 0.25 = 25\%$.

2. 28%. The probability of snow is 72%, so the probability that it will *not*
   snow is $100\% - 72\% = 28\%$.

3. $\dfrac{1}{8}$. Each time the coin is flipped, there is a $\dfrac{1}{2}$ probability for tails and
   a $\dfrac{1}{2}$ probability for heads. Getting tails, then heads, then tails

   is $\dfrac{1}{2} \times \dfrac{1}{2} \times \dfrac{1}{2} = \dfrac{1}{8}$.

4. $\frac{1}{5}$. Drawing a blue marble the first time is $\frac{3}{6}$ or $\frac{1}{2}$, but since she doesn't put the marble back, now there are only five marbles total and only two blue ones. The probability of getting a blue marble on the second draw is $\frac{2}{5}$. The probability of both of those occurring is $\frac{1}{2} \times \frac{2}{5} = \frac{1}{5}$.

EXERCISE 10-5

1. 63%. Since 47% own a car and 16% own a truck, we add those together to get 63%.

2. $\frac{7}{23}$. Be careful: We are being asked about rain when the chart shows days without rain. There are 16 of 31 days in August with no rain, so there are 15 with rain. There are 22 of 31 days in July with no rain, so there are 9 with rain. There are 26 of 30 days in July with no rain, so there are 4 with rain. Add up all the days with rain: $15 + 9 + 4 = 28$. Add up the total number of days: $31 + 31 + 30 = 92$. The probability of a summer day having rain is $\frac{28}{92} = \frac{7}{23}$.

3. $\frac{11}{43}$. This will include all the values with stems 3 and 4. Count them: 11. Count all the possible outcomes: 43. The probability that a randomly chosen number from the data set will be between 30 and 50 is $\frac{11}{43}$.

4. $\frac{2}{9}$

	apple
GC	banana
	yogurt
	apple
S	banana
	yogurt

(*continued*)

CS	apple
	banana
	yogurt

There are nine possible outcomes. Two of them result in a salad with a piece of fruit: salad/apple and salad/banana. The probability is $\frac{2}{9}$.

EXERCISE 10-6

1. 0.74, or 74%. Add up the probability of choosing a stuffed animal (0.41) and the probability of choosing a doll (0.33): $0.41 + 0.33 = 0.74$.

2.

Type of Home	Number of Children	Probability
single-family home	12	0.4
apartment	10	0.33
townhouse	8	0.27
total	30	1

EXERCISE 10-7

1. 5,040. Since order matters, all we need to do is multiply the number of choices he has for each of the four spots: 10 choices, then 9, then 8, then 7. $10 \times 9 \times 8 \times 7 = 5{,}040$.

2. 40,320. Since order matters, all we need to do is multiply the number of choices she has for each of the eight spots: $8 \times 7 \times 6 \times 5 \times 4 \times 3 \times 2 \times 1 = 40{,}320$. This multiplication is easier if we rearrange the numbers: $8 \times 5 = 40$, $40 \times 6 = 240$, $240 \times 6 = 960$, $960 \times 3 = 2{,}880$, $2{,}880 \times 2 = 5{,}760$, and $5{,}760 \times 7 = 40{,}320$.

3. 1,140. Since order doesn't matter, we need to find the permutations and then *divide* by the number of ways to arrange those three people. First, find the number of choices for each of the three spots: 20, then 19, then 18. Now find the number of ways to arrange those three spots: $3 \times 2 \times 1$.

We can reduce before we multiply: $\dfrac{20 \times 19 \times 18}{3 \times 2 \times 1} = \dfrac{20 \times 19 \times 3}{1} = 1{,}140$.

4. 15. Since order doesn't matter, we need to find the permutations and then *divide* by the number of ways to arrange a pair of people. First, find the number of choices for each of the two spots: 6, then 5. Now find the number of ways to arrange those two spots: 2×1. We can reduce before we multiply: $\dfrac{6 \times 5}{2 \times 1} = \dfrac{3 \times 5}{1} = 15$.

Geometry Fundamentals

EXERCISE 11-1

1. Point E is at $(-5, 2)$.

2. 5. There are five labeled points: A, B, C, D, and E.

3. $\overrightarrow{AB}$. One ray is shown: $\overrightarrow{AB}$. It cannot be written as $\overrightarrow{BA}$ because A is the vertex of the ray.

4. Line CD or line DC. Either is correct.

5. Any two of the following: $\overline{CD}$, $\overline{DC}$, $\overline{AB}$, $\overline{BA}$.

EXERCISE 11-2

1. $17°$; acute. $90° - 73° = 17°$.

2. $90°$; right. $180° - 90° = 90°$.

3. $135°$; obtuse. $180° - 45° = 135°$.

4. $22°$; acute. $90° - 68° = 22°$.

5. $90°$; right. The little box in the corner tells us this is a right angle, which is $90°$.

6. $62°$; acute. $180° - 118° = 62°$.

EXERCISE 11-3

1. $\angle a = 40°$ because it is a supplementary angle with $140°$.
   $180° - 140° = 40°$.

2. $\angle b = 140°$ because it is a vertical angle with $140°$ and vertical angles are equal.

3. $\angle c = 40°$ because it is a supplementary angle with $140°$.
   $180° - 140° = 40°$.

4. $\angle x = 140°$ because it is a supplementary angle with $40°$.
   $180° - 40° = 140°$.

5. $\angle y = 40°$ because it is a vertical angle with $40°$ and vertical angles are equal.

6. $\angle z = 140°$ because it is a supplementary angle with $40°$.
   $180° - 40° = 140°$.

7. $180°$ because $\angle b$ and $\angle y$ are consecutive interior angles, which always add up to $180°$.

8. $180°$ because $\angle c$ and $\angle z$ are consecutive interior angles, which always add up to $180°$.

9. $180°$ because any big angle plus any small angle add up to $180°$.

Geometric Figures

EXERCISE 12-1

1. a. equilateral; b. right; c. isosceles; d. scalene; e. equilateral

2. a. acute; b. right; c. obtuse; d. acute

3. a. 40° because this is an isosceles triangle and the angles opposite the equal sides are equal.

   b. 45° because the angles of a triangle add up to 180°.
   $2x + x + x = 180°$, so $4x = 180°$ and $x = 45°$

   c. 33°, because the angles of a triangle add up to 180°.
   $23° + 124° + x = 180°$ so $147° + x = 180°$ and $x = 33°$

4. a. $4^2 = 4$ because this is an equilateral triangle and all the sides are equal.

   b. $\overline{AC} = 4\sqrt{2}$. The area of triangle b is 8. Use the Pythagorean theorem to find $\overline{AC}$.

   $$4^2 + 4^2 = (\overline{AC})^2$$
   $$16 + 16 = (\overline{AC})^2$$
   $$32 = \overline{AC}^2$$
   $$\overline{AC} = \sqrt{32}$$
   $$= \sqrt{16 \times 2} = 4\sqrt{2}$$

   The area of a triangle is $A = \dfrac{1}{2}bh$, so $A = \dfrac{1}{2}(4)(4) = 8$.

5. Range of possible values for $\overline{AB}$: $9 < \overline{AB} < 19$. The third side of a triangle must be less than the sum of the other two sides and greater than the difference between the other two sides.

   Range of possible values for the perimeter: $28 < P < 38$. Once we have found the range for $\overline{AB}$, we can add that range to the known perimeter of the other two sides: 19.

6. The area is 11.75, and the perimeter is 18.95. The area of a triangle is

   $A = \dfrac{1}{2}bh$, so $A = \dfrac{1}{2}(7.65)(3.07) = \dfrac{1}{2}(23.5) = 11.75$. The perimeter is

   the sum of the sides: $3.07 + 7.65 + 8.23 = 18.95$.

7. $5x$. The perimeter is the sum of the sides: $x + x + 3x = 5x$.

EXERCISE 12-2

1. a. Square, 90°. A square has four equal sides and four 90° angles.

   b. Rhombus, 100°. A rhombus has four equal sides, opposite angles equal, and adjacent angles are supplementary. If $\angle D = 80°$, and $\angle D + x = 180°$, then $x = 100°$.

   c. Trapezoid, 143°. The angles of a quadrilateral must add up to 360°: $55° + 37° + 125° = 217°$ and $360° - 217° = 143°$.

   d. Kite, 110°. The angles between the equal sides are equal, so if the angle opposite $x = 110°$ then $x = 110°$. Also, the angles of a quadrilateral must add up to 360°. $360° - 60° - 80° - 110° = 110°$.

   e. Rectangle, 90°. A rectangle has equal opposite sides and four 90° angles.

   f. Parallelogram, 127.24°. A parallelogram has equal opposite sides, equal opposite angles, and supplementary adjacent angles. The angle opposite 52.76° is also 52.76°: $X + 52.76° = 180°$, so $x = 127.24°$.

2. a. $12.69\,\text{mm}^2$. The area of a rectangle is $A = lw$, so
   $$A = (5.93\,\text{mm})(2.14\,\text{mm}) = 12.69\,\text{mm}^2.$$

   b. $28\,\text{units}^2$. The area of a parallelogram is $A = bh$,
   so $A = (7\,\text{units})(4\,\text{units}) = 28\,\text{units}^2$

   c. $17.5\,\text{in}^2$. The area of a trapezoid is the average of the top and bottom times the height:
   $$A = \left(\frac{3\,\text{in} + 4\,\text{in}}{2}\right)(5\,\text{in}) = \left(\frac{7}{2}\,\text{in}\right)(5\,\text{in}) = \frac{35}{2}\,\text{in}^2 = 17.5\,\text{in}^2$$

   d. $39.69\,\text{in}^2$. The area of a square is $A = s^2$, so
   $$A = (6.3\,\text{in})^2 = 39.69\,\text{in}^2$$

   e. $22\,\text{cm}^2$. The area of a kite is the product of the diagonals divided by two: $A = \dfrac{(11\,\text{cm})(4\,\text{cm})}{2} = \dfrac{44\,\text{cm}^2}{2} = 22\,\text{cm}^2$.

   f. $112\,\text{in}^2$. Divide the figure into a rectangle and a triangle, like this:

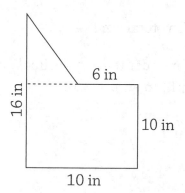

Now we have a 10-in square and a triangle with base 4 in $(10 - 6)$ and height 6 in $(16 - 10)$. The area of the square is $A = s^2$,

so $A = (10\,\text{in})^2 = 100\,\text{in}^2$. The area of the triangle is $A = \dfrac{1}{2}bh$, so

$A = \dfrac{1}{2}(4\,\text{in})(6\,\text{in}) = 12\,\text{in}^2$. Add those together to get the total

area: $100\,\text{in} + 12\,\text{in}^2 = 112\,\text{in}^2$.

EXERCISE 12-3

1. $46.3°$. The diameter forms a $180°$ angle, and the marked angles are $88.5°$ and $45.2°$, so we can subtract those from $180°$ to find x.

   $180° - 88.5° - 45.2° = 46.3°$.

2. $7\pi\,\text{cm}$. $C = 2\pi r$, so $C = 2\pi(3.5\,\text{cm}) = 7\pi\,\text{cm}$.

3. $25\pi\,\text{cm}^2$. $A = \pi r^2$. The diameter is 10 cm, so the radius is 5 cm. $A = \pi(5\,\text{cm})^2 = 25\pi\,\text{cm}^2$.

4. 6. $A = \pi r^2$, so $9\pi = \pi r^2$. Divide both sides by π: $\dfrac{9\pi}{\pi} = \dfrac{\pi r^2}{\pi}$, to get

   $9 = r^2$. Take the square root of both sides to get $r = \sqrt{9} = 3$. Double that to find the diameter: $3 \times 2 = 6$.

5. $189.3\,\text{in}^2$. The figure is a rectangle plus a semicircle. The area of the rectangle is $A = lw$, so $A = (15\,\text{in})(10\,\text{in}) = 150\,\text{in}^2$. The area of the

   semicircle is half the area of the circle, so $A = \dfrac{1}{2}\pi r^2$. The diameter

   of the semicircle is 10 in, so the radius is 5 in.

$A = \frac{1}{2}\pi(5\,\text{in})^2 = \frac{1}{2}\pi(25\,\text{in}^2) = 12.5\,\pi\,\text{in}^2$. The total area is

$150\,\text{in}^2 + 12.5\pi\,\text{in}^2$. We need to estimate π in order to get a decimal answer that we can round to the nearest tenth, so we use 3.14 for π.
$(12.5)(3.14)\,\text{in}^2 = 39.25\,\text{in}^2$.
$150\,\text{in}^2 + 39.25\,\text{in}^2 = 189.25\,\text{in}^2 = 189.3\,\text{in}^2$.

EXERCISE 12-4

1. $57.66\,\text{cm}^2$. $S_A = 6s^2$, so $S_A = 6(3.1\,\text{cm})^2 = (6)(9.61\,\text{cm}^2) = 57.66\,\text{cm}^2$.

2. We need to find the area of each side and then add them all up. Two sides are 4 inches by 5 inches, so their area is $4 \times 5 = 20\,\text{in}^2$ each. Two sides are 5 inches by 7 inches, so their area is $5 \times 7 = 35\,\text{in}^2$ each. Two sides are 4 inches by 7 inches, so their area is $4 \times 7 = 28\,\text{in}^2$ each. Add those all up to get:
$20\,\text{in}^2 + 20\,\text{in}^2 + 35\,\text{in}^2 + 35\,\text{in}^2 + 28\,\text{in}^2 + 28\,\text{in}^2 = 166\,\text{in}^2$.

3. $360\,\text{cm}^2$. We need to find the area of each side and then add them all up. Two sides are triangles with base 12 cm and height 5 cm. Their area is

   $A = \frac{1}{2}bh$, so $A = \frac{1}{2}(12\,\text{cm})(5\,\text{cm}) = 30\,\text{cm}^2$ each. We also have three

   rectangles. One is 12 cm by 10 cm, so its area is $12 \times 10 = 120\,\text{cm}^2$.
   One is 13 cm by 10 cm, so its area is $13 \times 10 = 130\,\text{cm}^2$. One is 5 cm by 10 cm, so its area is $5 \times 10 = 50\,\text{cm}^2$. Add all the areas up to get:
   $30\,\text{cm}^2 + 30\,\text{cm}^2 + 120\,\text{cm}^2 + 130\,\text{cm}^2 + 50\,\text{cm}^2 = 360\,\text{cm}^2$.

4. 66π. To find the surface area, we need the area of the two circular ends, plus the area of the rectangle that forms the cylinder. The circle has a radius of 3, so $A = \pi(3)^2 = 9\pi$. The area of the rectangle is 8 times the circumference, and $C = 2\pi r$, so the rectangle is $2\pi(3) \times 8 = 6\pi \times 8 = 48\pi$. Add those together to get: $9\pi + 9\pi + 48\pi = 66\pi$.

EXERCISE 12-5

1. $13.824\,\text{in}^3$. $V = s^3$, and each side measures 2.4 inches,
   so $V = (2.4)^3 = 13.824\,\text{in}^3$.

2. $39\,\text{mm}^3$. $V = l \times w \times h$, so $V = (3\,\text{mm})(6.5\,\text{mm})(2\,\text{mm}) = 39\,\text{mm}^3$.

3. $300\,\text{cm}^3$. The base of the triangle is 12 centimeters, and the height is
   5 centimeters. The length of the prism is 10 centimeters: $V = \dfrac{1}{2}bhl$,
   so $V = \dfrac{1}{2}(12\,\text{cm})(5\,\text{cm})(10\,\text{cm}) = 300\,\text{cm}^3$.

4. $25{,}000\pi\,\text{in}^3$. The radius is 25 inches, and the height is 40 inches.
   $V = \pi r^2 h$, so $V = \pi(25\,\text{in.})^2(40\,\text{in}) = (625\pi\,\text{in}^2)(40\,\text{in}) = 25000\pi\,\text{in}^3$.

5. $57.2\pi\,\text{cm}^3$. The radius is 3.5 cm. $V = \dfrac{4}{3}\pi r^3$,
   so $V = \dfrac{4}{3}\pi(3.5\,\text{cm})^3 = 57.2\pi\,\text{cm}^3$.

6. $3.75\,\text{in}^3$. The radius is 1.5 inches, and the height is 5 inches: $V = \dfrac{1}{3}\pi r^2 h$,
   so $V = \dfrac{1}{3}\pi(1.5\,\text{in})^2(5\,\text{in}) = 3.75\,\text{in}^3$.

7. $4\,\text{cm}^3$. The length is 2 cm, the width is also 2 cm, and the height is 3 cm.
   $V = \dfrac{lwh}{3}$, so $V = \dfrac{(2\,\text{cm})(2\,\text{cm})(3\,\text{cm})}{3} = \dfrac{12\,\text{cm}^3}{3} = 4\,\text{cm}^3$.

EXERCISE 12-6

1. 2 in by 3 in. An image that is 4 inches by 6 inches, reduced by a scale
   factor of $\dfrac{1}{2}$, means the new image will be $(4\,\text{in}) \times \left(\dfrac{1}{2}\right)$ by $(6\,\text{in}) \times \left(\dfrac{1}{2}\right)$,
   or 2 in by 3 in.

2.

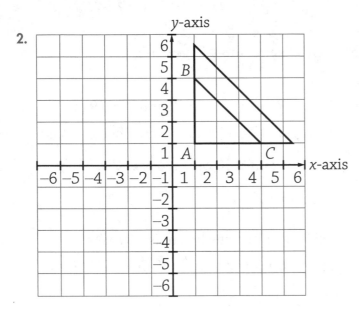

The original triangle has a base of 3 and a height of 3. Enlarging it by a scale factor of 1.5 means the new triangle will have a base of $(3)(1.5) = 4.5$ and a height of $(3)(1.5) = 4.5$.

3. The first and third figures are congruent.

4. 29°. The angles of a triangle must add to 180°. For $\triangle ABC$, one angle is 90°, and the other is 61°, so $\angle BAC$ is $180° - 90° - 61° = 29°$. $\angle EDF$ $\angle BAC$ is congruent to $\angle EDF$, so $\angle EDF = 29°$.

5. One. The figure has only one line of symmetry:

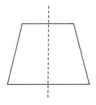

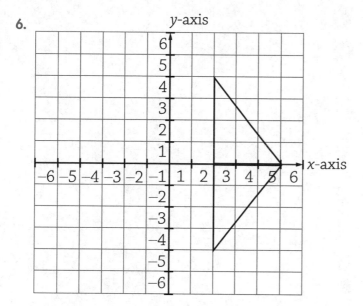

6.